T0259124

Computability Theory

Computability Theory
An Introduction to Recursion Theory

Herbert B. Enderton
University of California, Los Angeles

ELSEVIER

AMSTERDAM • BOSTON • HEIDELBERG • LONDON
NEW YORK • OXFORD • PARIS • SAN DIEGO
SAN FRANCISCO • SINGAPORE • SYDNEY • TOKYO
Academic Press is an imprint of Elsevier

Academic Press is an imprint of Elsevier
30 Corporate Drive, Suite 400, Burlington, MA 01803, USA
525 B Street, Suite 1800, San Diego, California 92101-4495, USA
84 Theobald's Road, London WC1X 8RR, UK

Notices
Knowledge and best practice in this field are constantly changing. As new research and experience
broaden our understanding, changes in research methods, professional practices, or medical treatment may
become necessary.

Practitioners and researchers must always rely on their own experience and knowledge in evaluating and
using any information, methods, compounds, or experiments described herein. In using such information
or methods they should be mindful of their own safety and the safety of others, including parties for whom
they have a professional responsibility.

To the fullest extent of the law, neither the Publisher nor the authors, contributors, or editors, assume any
liability for any injury and/or damage to persons or property as a matter of products liability, negligence or
otherwise, or from any use or operation of any methods, products, instructions, or ideas contained in the
material herein.

Library of Congress Cataloging-in-Publication Data
Enderton, Herbert B.
  Computability theory : an introduction to recursion theory / Herbert B. Enderton.
    p. cm.
  ISBN 978-0-12-384958-8 (hardback)
  1. Recursion theory. I. Title.
  QA9.6.E53 2011
  511.3'5–dc22
                                    2010038448
British Library Cataloguing-in-Publication Data
A catalogue record for this book is available from the British Library.

ISBN: 978-0-12-384958-8

For information on all Academic Press publications
visit our Web site at *www.elsevierdirect.com*

Typeset by: diacriTech, India

for Cathy

Contents

Foreword

The study of the class of computable partial functions (i.e., recursive partial functions) stands at the intersection of three fields: mathematics, theoretical computer science, and philosophy.

- Mathematically, computability theory originates from the concept of an *algorithm*. It leads to a classification of functions according their inherent *complexity*.
- For the computer scientist, computability theory shows that quite apart from practical matters of running time and memory space, there is a purely theoretical limit to what computer programs can do. This is an important fact, and leads to the questions: Where is the limit? What is on this side of the limit, and what lies beyond it?
- Computability is relevant to the philosophy of mathematics and, in particular, to the questions: What *is* a proof? Does every true sentence have a proof?

Computability theory is not an ancient branch of mathematics; it started in 1936. In that year, Alonzo Church, Alan Turing, and Emil Post each published fundamental papers that characterized the class of computable partial functions. Church's article introduced what is now called "Church's thesis" (or the Church–Turing thesis), to be discussed in Chapter 1. Turing's article introduced what are now called "Turing machines." (1936 was also the year in which *The Journal of Symbolic Logic* began publication, under the leadership of Alonzo Church and others. Finally, it was also the year in which I was born.)

Preface

This book is intended to serve as a textbook for a one-term course on computability theory (i.e., recursion theory), for upper-division mathematics and computer science students. And the book is *focused* on this one topic, to the exclusion of such computer-science topics as automata theory, context-free languages, and the like. This makes it possible to get fairly quickly to core results in computability theory, such as the unsolvability of the halting problem.

The only prerequisite for reading this book is a willingness to tolerate a certain level of abstraction and rigor. The goal here is to prove theorems, not to calculate numbers or write computer programs. The book uses standard mathematical jargon; there is an appendix on "Mathspeak" to explain some of this jargon.

The basic material is covered in Chapters 1–4. After reading those chapters, Chapters 5, 6, and 7, which are largely independent of each other, can be read in any order.

Chapter 1 is an informal introduction to the concepts of computability theory. That is, instead of an emphasis on precise definitions and rigorous proofs, the goal is to convey an intuitive understanding of the basic concepts. The precision and rigor will be in the later chapters; first one needs an insight into the nature of the concepts.

Chapters 2 and 3 explore two of the ways in which the concept of effective calculability can be made precise. The two ways are proven to be equivalent. This allows the definition of "computable partial function" to be made on the basis of the common ground. (The alternative phrase, "recursive partial function," is kept in the background.)

The interplay between the approaches of Chapters 2 and 3 yields proofs of such basic results as the unsolvability of the halting problem. The interplay also yields a proof of the enumeration theorem (without appealing to the reader's experience with high level programming languages).

Chapter 4 presents the properties of recursively enumerable (r.e.) sets. (Note the shift in terminology here; this time, the phrase "computably enumerable" is kept in the background. The hope is that the reader will emerge being bilingual.)

Chapter 5 connects computability theory to the Gödel incompleteness theorem. The heart of the incompleteness theorem lies in the fact that the set of Gödel numbers of true sentences of arithmetic is a set that is "productive" in the sense defined by Emil Post. And this fact is squarely within the domain of computability theory.

Chapter 6 introduces relative computability and the degrees of unsolvability.

Chapter 7 introduces polynomial time computability and discusses the "P versus NP" problem. In this final chapter, not everything receives a complete proof. Instead, the intent is to give a transition to a later study of computational complexity.

1 The Computability Concept

1.1 The Informal Concept

1.1.1 Decidable Sets

Computability theory, also known as recursion theory, is the area of mathematics dealing with the concept of an *effective procedure* – a procedure that can be carried out by following specific rules. For example, we might ask whether there is some effective procedure – some algorithm – that, given a sentence about the integers, will decide whether that sentence is true or false. In other words, is the set of true sentences about the integers *decidable*? (We will see later that the answer is negative.) Or for a simpler example, the set of prime numbers is certainly a decidable set. That is, there are quite mechanical procedures, which are taught in the schools, for deciding whether or not any given integer is a prime number. (For a very large number, the procedure taught in the schools might take a long time.) If we want, we can write a computer program to execute the procedure. Simpler still, the set of even integers is decidable. We can write a computer program that, given an integer, will very quickly decide whether or not it is even. Our goal is to study what decision problems can be solved (in principle) by a computer program, and what decision problems (if any) cannot.

More generally, consider a set S of natural numbers. (The natural numbers are $0, 1, 2, \ldots$. In particular, 0 is natural.) We say that S is a *decidable* set if there exists an effective procedure that, given any natural number, will eventually end by supplying us with the answer. "Yes" if the given number is a member of S and "No" if it is not a member of S.

(Initially, we are going to examine computability in the context of the natural numbers. Later, we will see that computability concepts can be readily transferred to the context of strings of letters from a finite alphabet. In that context, we can consider a set S of strings, such as the set of equations, like $x(y + z) = xy + xz$ that hold in the algebra of real numbers. But to start with, we will consider sets of natural numbers.)

And by an *effective procedure* here is meant a procedure for which we can give exact instructions – a program – for carrying out the procedure. Following these instructions should not demand brilliant insights on the part of the agent (human or machine) following them. It must be possible, at least in principle, to make the instructions so explicit that they can be executed by a diligent clerk (who is very good at following directions but is not too clever) or even a machine (which does not think at all). That is, it must be possible for our instructions to be *mechanically implemented*. (One might imagine a mathematician so brilliant that he or she can look at any sentence of arithmetic and say whether it is true or false. But you cannot ask the clerk to do this. And there is no computer program to do this. It is not merely that we have not succeeded in writing such a program. We can actually prove that such a program cannot possibly exist!)

Although these instructions must, of course, be finite in length, we impose no upper bound on their possible length. We do not rule out the possibility that the instructions might even be absurdly long. (If the number of lines in the instructions exceeds the number of electrons in the universe, we merely shrug and say, "That's a pretty long program.") We insist only that the instructions – the program – be finitely long, so that we can *communicate* them to the person or machine doing the calculations. (There is no way to give someone all of an infinite object.) Similarly, in order to obtain the most comprehensive concepts, we impose no bounds on the time that the procedure might consume before it supplies us with the answer. Nor do we impose a bound on the amount of storage space (scratch paper) that the procedure might need to use. (The procedure might, for example, need to utilize very large numbers requiring a substantial amount of space simply to write down.) We merely insist that the procedure give us the answer eventually, in some finite length of time. What is definitely ruled out is doing infinitely many steps and *then* giving the answer.

In Chapter 7, we will consider more restrictive concepts, where the amount of time is limited in some way, so as to exclude the possibility of ridiculously long execution times. But initially, we want to avoid such restrictions to obtain the limiting case where practical limitations on execution time or memory space are removed. It is well known that in the real world, the speed and capability of computers has been steadily growing. We want to ignore actual speed and actual capability, and instead we want to ask what the purely theoretical limits are.

The foregoing description of effective procedures is admittedly vague and imprecise. In the following section, we will look at how this vague description can be made precise – how the concept can be made into a *mathematical* concept. Nonetheless, the informal idea of what can be done by effective procedure, that is, what is *calculable*, can be very useful. Rigor and precision can wait until the *next* chapter. First we need a sense of where we are going.

For example, any finite set of natural numbers must be decidable. The program for the decision procedure can simply include a list of all the numbers in the set. Then given a number, the program can check it against the list. Thus, the concept of decidability is interesting only for infinite sets.

Our description of effective procedures, vague as it is, already shows how limiting the concept of decidability is. One can, for example, utilize the concepts of countable and uncountable sets (see the appendix for a summary of these concepts). It is not hard to see that there are only countably many possible instructions of finite length that one can write out (using a standard keyboard, say). But there are uncountably many sets of natural numbers (by Cantor's diagonal argument). It follows that almost all sets, in a sense, are *undecidable*.

The fact that not every set is decidable is relevant to theoretical computer science. The fact that there is a limit to what can be carried out by effective procedures means that there is a limit to what can – even in principle – be done by computer programs. And this raises the questions: What can be done? What cannot?

Historically, computability theory arose before the development of digital computers. Computability theory is relevant to certain considerations in mathematical logic. At the heart of mathematical activity is the proving of theorems. Consider what is required for a string of symbols to constitute an "acceptable mathematical proof." Before we accept a proof, and add the result being proved to our storehouse of mathematical knowledge, we insist that the proof be *verifiable*. That is, it should be possible for another mathematician, such as the referee of the article containing the proof, to check, step by step, the correctness of the proof. Eventually, the referee either concludes that the proof is indeed correct or concludes that the proof contains a gap or an error and is not yet acceptable. That is, the set of acceptable mathematical proofs – regarded as strings of symbols – should be *decidable*. This fact will be seen (in a later chapter) to have significant consequences for what can and cannot be proved. We conclude that computability theory is relevant to the foundations of mathematics. But if logicians had not invented the computability concept, then computer scientists would later have done so.

1.1.2 Calculable Functions

Before going on, we should broaden the canvas from considering decidable and undecidable sets to considering the more general situation of *partial functions*. Let $\mathbb{N} = \{0, 1, 2, \ldots\}$ be the set of natural numbers. Then, an example of a two-place function on $\mathbb{N}$ is the subtraction function

$$g(m, n) = \begin{cases} m - n & \text{if } m \geq n \\ 0 & \text{otherwise} \end{cases}$$

(where we have avoided negative numbers). A different subtraction function is the "partial" function

$$f(m, n) = \begin{cases} m - n & \text{if } m \geq n \\ \uparrow & \text{otherwise} \end{cases}$$

where "$\uparrow$" indicates that the function is undefined. Thus $f(5, 2) = 3$, but $f(2, 5)$ is undefined; the pair $\langle 2, 5 \rangle$ is not in the domain of f.

In general, say that a k-place *partial function* on $\mathbb{N}$ is a function whose domain is some set of k-tuples of natural numbers and whose values are natural numbers. In other words, for a k-place partial function f and a k-tuple $\langle x_1, \ldots, x_k \rangle$, possibly $f(x_1, \ldots, x_k)$ is defined (i.e., $\langle x_1, \ldots, x_k \rangle$ is in the domain of f), in which case the function value $f(x_1, \ldots, x_k)$ is in $\mathbb{N}$, and possibly $f(x_1, \ldots, x_k)$ is undefined (i.e., $\langle x_1, \ldots, x_k \rangle$ is not in the domain of f).

At one extreme, there are partial functions whose domains are the set $\mathbb{N}^k$ of *all* k-tuples; such functions are said to be *total*. (The adjective "partial" covers both the total and the nontotal functions.) At the other extreme, there is the empty function, that is, the function that is defined nowhere. The empty function might not seem particularly useful, but it does count as one of the k-place partial functions.

For a k-place partial function f, we say that f is an *effectively calculable partial function* if there exists an effective procedure with the following property:

- Given a k-tuple $\vec{x}$ in the domain of f, the procedure eventually halts and returns the correct value for $f(\vec{x})$.
- Given a k-tuple $\vec{x}$ *not* in the domain of f, the procedure does not halt and return a value.

(There is one issue here: How can a number be *given*? To communicate a number x to the procedure, we send it the *numeral* for x. Numerals are bits of language, which can be communicated. Numbers are not. Communication requires language. Nonetheless, we will continue to speak of being "given numbers m and n" and so forth. But at a few points, we will need to be more accurate and take account of the fact that what the procedure is given are numerals. There was a time in the 1960s when, as part of the "new math," schoolteachers were encouraged to distinguish carefully between numbers and numerals. This was a good idea that turned out not to work.)

For example, the partial function for subtraction

$$f(m, n) = \begin{cases} m - n & \text{if } m \geq n \\ \uparrow & \text{otherwise} \end{cases}$$

is effectively calculable, and procedures for calculating it, using base-10 numerals, are taught in the elementary schools.

The empty function is effectively calculable. The effective procedure for it, given a k-tuple, does not need to do anything in particular. But it must never halt and return a value.

The concept of decidability can then be described in terms of functions: For a subset S of $\mathbb{N}^k$, we can say that S is *decidable* iff its characteristic function

$$C_S(\vec{x}) = \begin{cases} \text{Yes} & \text{if } \vec{x} \in S \\ \text{No} & \text{if } \vec{x} \notin S \end{cases}$$

(which is always total) is effectively calculable. Here "Yes" and "No" are some fixed members of $\mathbb{N}$, such as 1 and 0.

(That word "iff" in the preceding paragraph means "if and only if." This is a bit of mathematical jargon that has proved to be so useful that it has become a standard part of mathspeak.)

Here, if $k = 1$, then S is a set of numbers. If $k = 2$, then we have the concept of a decidable binary relation on numbers, and so forth. Take, for example, the divisibility relation, that is, the set of pairs $\langle m, n \rangle$ such that m divides n evenly. (For definiteness, assume that 0 divides only itself.) The divisibility relation is decidable because given m and n, we can carry out the division algorithm we all learned in the fourth grade, and see whether the remainder is 0 or not.

Example: Any total constant function on $\mathbb{N}$ is effectively calculable. Suppose, for example, $f(x) = 36$ for all x in $\mathbb{N}$. There is an obvious procedure for calculating f; it ignores its input and writes "36" as the output. This may seem a triviality, but compare it with the next example.

Example: Define the function F as follows.

$$F(x) = \begin{cases} 1 & \text{if Goldbach's conjecture is true} \\ 0 & \text{if Goldbach's conjecture is false} \end{cases}$$

Goldbach's conjecture states that every even integer greater than 2 is the sum of two primes; for example, $22 = 5 + 17$. This conjecture is still an open problem in mathematics. Is this function F effectively calculable? (Choose your answer before reading the next paragraph.)

Observe that F is a total constant function. (Classical logic enters here: Either there is an even number that serves as a counterexample or there isn't.) So as noted in the preceding example, F is effectively calculable. What, then, is a procedure for computing F? I don't know, but I can give you two procedures and be confident that one of them computes F.

The point of this example is that effective calculability is a property of the function itself, not a property of some linguistic description used to specify the function. (One says that the effective calculability property is *extensional*.) There are many English phrases that would serve to define F. For a function to be effectively calculable, there must *exist* (in the mathematical sense) an effective procedure for computing it. That is not the same as saying that you hold such a procedure in your hand. If, in the year 2083, some creature in the universe proves (or refutes) Goldbach's conjecture, then it does *not* mean that F will suddenly change from noncalculable to calculable. It was calculable all along.

There will be, however, situations later in which we will want more than the mere existence of an effective procedure P; we will want some way of actually finding P, given some suitable clues. That is for later.

It is very natural to extend these concepts to the situation where we have half of decidability: Say that S is *semidecidable* if its "semicharacteristic function"

$$c_S(\vec{x}) = \begin{cases} \text{Yes} & \text{if } \vec{x} \in S \\ \uparrow & \text{if } \vec{x} \notin S \end{cases}$$

is an effectively calculable partial function. Thus, a set S of numbers is semidecidable if there is an effective procedure for *recognizing* members of S. We can think of S as the set that the procedure *accepts*. And the effective procedure, while it may not be a decision procedure, is at least an *acceptance* procedure.

Any decidable set is also semidecidable. If we have an effective procedure that calculates the characteristic function C_S, then we can convert it to an effective procedure that calculates the semicharacteristic function c_S. We simply replace each "output No" command by some endless loop. Or more informally, we simply unscrew the No bulb.

What about the converse? Are there semidecidable sets that are not decidable? We will see that there are indeed. The trouble with the semicharacteristic function is that it never produces a No answer. Suppose that we have been calculating $c_S(\vec{x})$ for 37 years, and the procedure has not yet terminated. Should we give up and conclude that $\vec{x}$ is not in S? Or maybe working just another ten minutes would yield the information that $\bar{x}$ does belong to S. There is, in general, no way to know.

Here is another example of a calculable partial function:

$$F(n) = \text{the smallest } p > n \text{ such that both } p \text{ and } p + 2 \text{ are prime}$$

Here it is to be understood that $F(n)$ is undefined if there is no number p as described; thus F might not be total. For example, $F(9) = 11$ because both 11 and 13 are prime. It is not known whether or not F is total. The "twin prime conjecture," which says that there are infinitely many pairs of primes that differ by 2, is equivalent to the statement that F is total. The twin prime conjecture is still an open problem. Nonetheless, we can be certain that F is effectively calculable. One procedure for calculating $F(n)$ proceeds as follows. "Given n, first put $p = n + 1$. Then check whether or not p and $p + 2$ are both prime. If they are, then stop and give output p. If not, increment p and continue." What if n is huge, say, $n = 10^{10^{10}}$? On the one hand, if there is a larger prime pair, then this procedure will find the first one, and halt with the correct output. On the other hand, if there is no larger prime pair, then the procedure never halts, so it never gives us an answer. That is all right; because $F(n)$ is undefined, the procedure *should not* give us any answer.

Now suppose we modify this example. Consider the total function:

$$G(n) = \begin{cases} F(n) & \text{if } F(n) \downarrow \\ 0 & \text{otherwise} \end{cases}$$

Here "$F(n) \downarrow$" means that $F(n)$ is defined, so that n belongs to the domain of F. Then the function G is *also* effectively calculable. That is, there *exists* a program that calculates G correctly.

The twin prime conjecture is either true or false: Either the prime pairs go on forever, or there is a largest one. (At this point, classical logic enters once again.) In the first case, $F = G$ and the effective procedure for F also computes G. In the second case, G is eventually constantly 0. And any eventually constant function is calculable (the procedure can utilize a table for the finite part of the function before it stabilizes).

So in either case, there *exists* an effective procedure for G. That is not the same as saying that we *know* that procedure. This example indicates once again the difference between knowing that a certain effective procedure exists and having the effective procedure in our hands (or having convincing reasons for knowing that the procedure in our hands will work).

One person's program is another person's data. This is the principle behind operating systems (and behind the idea of a stored-program computer). One's favorite program is, to the operating system, another piece of data to be received as input and processed. The operating system is calculating the values of a two-place "universal" function. We next want to see if these concepts can be applied to our study of calculable functions. (Historically, the flow of ideas was in exactly the opposite direction! The following digression expands on this point.)

Digression: The concept of a general-purpose, stored-program computer is now very common, but the concept developed slowly over a period of time. The ENIAC machine, the most important computer of the 1940s, was programmed by setting switches and inserting cables into plugboards! This is a far cry from treating a program like data. It was von Neumann who, in a 1945 technical report, laid out the crucial ideas for a general-purpose stored-program computer, that is, for a universal computer. Turing's 1936 article on what are now called Turing machines had proved the existence of a "universal Turing machine" to compute the Φ function described below. When Turing went to Princeton in 1936–37, von Neumann was there and must have been aware of his work. Apparently, von Neumann's thinking in 1945 was influenced by Turing's work of nearly a decade earlier.

Suppose we adopt a fixed method of encoding any set of instructions by a single natural number. (First, we convert the instructions to a string of 0's and 1's – one always does this with computer programs – and then we regard that string as naming a natural number under a suitable base-2 notation.) Then the "universal function"

$\Phi(w, x) = $ the result of applying the instructions coded by w to the input x

is an effectively calculable partial function (where it is understood that $\Phi(w, x)$ is undefined whenever applying the instructions coded by w to the input x fails to halt and return an output). Here are the instructions for Φ: "Given w and x, decode w to see what it says to do with x, and then do it." Of course, the function Φ is not total. For one thing, when we try to decode w, we might get complete nonsense, so that the instruction "then do it" leads nowhere. And even if decoding w yields explicit and comprehensible instructions, applying those instructions to a particular x might never yield an output.

(The reasoning here will be repeated in Chapter 3, when we will have more concrete material to deal with. But the guiding ideas will remain the same.)

The two-place partial function Φ is "universal" in the sense that *any* one-place effectively calculable partial function f is given by the equation

$f(x) = \Phi(e, x) \quad$ for all x

where e codes the instructions for f. It will be helpful to introduce a special notation here: Let $[\![e]\!]$ be the one-place partial function defined by the equation

$[\![e]\!](x) = \Phi(e, x).$

That is, $[\![e]\!]$ is the partial function whose instructions are coded by e, with the understanding that, because some values of e might not code anything sensible, the function $[\![e]\!]$ might be the empty function. In any case, $[\![e]\!]$ is the partial function we get from Φ, when we hold its first variable fixed at e. Thus,

$[\![0]\!], [\![1]\!], [\![2]\!], \ldots$

is a complete list (with repetitions) of all the one-place effectively calculable partial functions. The values of $[\![e]\!]$ are given by the $(e + 1)$st row in the following table:

$[\![0]\!]$	$\Phi(0, 0)$	$\Phi(0, 1)$	$\Phi(0, 2)$	$\Phi(0, 3)$	$\cdots$
$[\![1]\!]$	$\Phi(1, 0)$	$\Phi(1, 1)$	$\Phi(1, 2)$	$\Phi(1, 3)$	$\cdots$
$[\![2]\!]$	$\Phi(2, 0)$	$\Phi(2, 1)$	$\Phi(2, 2)$	$\Phi(2, 3)$	$\cdots$
$[\![3]\!]$	$\Phi(3, 0)$	$\Phi(3, 1)$	$\Phi(3, 2)$	$\Phi(3, 3)$	$\cdots$
$\cdots$	$\cdots$	$\cdots$	$\cdots$	$\cdots$	

Using the universal partial function Φ, we can construct an *undecidable* binary relation, the *halting* relation H:

$$\langle w, x \rangle \in H \iff \Phi(w, x) \downarrow$$
$$\iff \text{applying the instructions coded by } w \text{ to input } x \text{ halts}$$

On the positive side, H is semidecidable. To calculate the semicharacteristic function $c_H(w, x)$, given w and x, we first calculate $\Phi(w, x)$. If and when this halts and returns a value, we give output "Yes" and stop.

On the negative side, H is not decidable. To see this, first consider the following partial function:

$$f(x) = \begin{cases} \text{Yes} & \text{if } \Phi(x, x) \uparrow \\ \uparrow & \text{if } \Phi(x, x) \downarrow \end{cases}$$

(Notice that we are using the classical diagonal construction. Looking at the earlier table of the values of Φ arranged in a two-dimensional array, one sees that f has been made by going along the diagonal of that table, taking the entry $\Phi(x, x)$ found there, and making sure that $f(x)$ differs from it.)

There are two things to be said about f. First, f cannot possibly be effectively calculable. Consider any set of instructions that *might* compute f. Those instructions have some code number k and hence compute the partial function $[\![k]\!]$. Could that be the same as f? No, f and $[\![k]\!]$ differ at the input k. That is, f has been constructed in such a way that $f(k)$ differs from $[\![k]\!](k)$; they differ because one is defined and the other is not. So these instructions cannot correctly compute f; they produce the wrong result at the input k. And because k was arbitrary, we are forced to conclude that *no* set of instructions can correctly compute f. (This is our first example of a partial function that is not effectively calculable. There are a great many more, as will be seen.)

Secondly, we can argue that *if* we had a decision procedure for H, then we *could* calculate f. To compute $f(x)$, we first use that decision procedure for H to decide if $(x, x) \in H$ or not. If not, then $f(x) = $ Yes. But if $(x, x) \in H$, then the procedure for finding $f(x)$ should throw itself into an infinite loop because $f(x)$ is undefined.

Putting these two observations about f together, we conclude that there can be no decision procedure for H. The fact that H is undecidable is usually expressed by saying that "the halting problem is unsolvable"; i.e., we cannot in general effectively determine, given w and x, whether applying the instructions coded by w to the input x will eventually terminate or will go on forever:

Unsolvability of the halting problem: *The relation*

$$\{\langle w, x \rangle \mid applying\ instructions\ coded\ by\ w\ to\ input\ x\ halts\}$$

is semidecidable but not decidable.

The function f in the preceding argument

$$f(x) = \begin{cases} \text{Yes} & \text{if } \Phi(x, x) \uparrow \\ \uparrow & \text{if } \Phi(x, x) \downarrow \end{cases}$$

is the semicharacteristic function of the set $\{x \mid \Phi(x, x) \uparrow\}$. Because its semicharacteristic function is not effectively calculable, we can conclude that this set is *not* semidecidable.

Let K be the complement of this set:

$$K = \{x \mid \Phi(x, x) \downarrow\} = \{x \mid [\![x]\!](x) \downarrow\}.$$

This set *is* semidecidable. How might we compute $c_K(x)$, given x? We try to compute $\Phi(x, x)$ (which is possible because Φ is an effectively calculable partial function). If and when the calculation halts and returns an output, we give the output "Yes" and stop. Until such time, we keep trying. (This argument is the same one that we saw for the semidecidability of H. And $x \in K \Leftrightarrow \langle x, x \rangle \in H$.)

Kleene's theorem: *A set (or a relation) is decidable if and only if both it and its complement are semidecidable.*

Here if we are working with sets of numbers, then the complement is with respect to $\mathbb{N}$; if we are working with a k-ary relation, then the complement is with respect to $\mathbb{N}^k$.

Proof. On the one hand, if a set S is decidable, then its complement $\overline{S}$ is also decidable – we simply switch the "Yes" and the "No." So both S and its complement $\overline{S}$ are semidecidable because decidable sets are also semidecidable.

On the other hand, suppose that S is a set for which both c_S and $c_{\overline{S}}$ are effectively calculable. The idea is to glue together these two halves of a decision procedure to make a whole one. Say we want to find $C_S(x)$, given x. We need to organize our time. During odd-numbered minutes, we run our program for $c_S(x)$. During even-numbered minutes, we run our program for $c_{\overline{S}}(x)$. Of course, at the end of each minute, we store away what we have done, so that we can later pick up from where we left off.

Eventually, we must receive a "Yes." If during an odd-numbered minute, we find that $c_S(x) = $ Yes (this must happen eventually if $x \in S$), then we give output "Yes" and stop. And if during an even-numbered minute, we find that $c_{\overline{S}}(x) = $ Yes (this must happen eventually if $x \notin S$), then we give output "No" and stop.

(Alternatively, we can imagine working ambidextrously. With the left hand, we work on calculating $c_S(x)$; with the right hand, we work on calculating $c_{\overline{S}}(x)$. Eventually, one hand discovers the answer.)　　　　　　　　　　　　　　　　　　　⊣

The set K is an example of a semidecidable set that is not decidable. Its complement $\overline{K}$ is not semidecidable; we have seen that its semicharacteristic function f is not effectively calculable.

The connection between effectively calculable partial functions and semidecidable sets can be further described as follows:

Theorem:

(i) *A relation is semidecidable if and only if it is the domain of some effectively calculable partial function.*

(ii) *A partial function f is an effectively calculable partial function if and only if its graph G (i.e., the set of tuples $\langle \vec{x}, y \rangle$ such that $f(\vec{x}) = y$) is a semidecidable relation.*

Proof. For statement (i), one direction is true by definition: Any relation is the domain of its semicharacteristic function, and for a semidecidable relation, that function is an effectively calculable partial function.

Conversely, for an effectively calculable partial function, f, we have the natural semidecision procedure for its domain: Given $\vec{x}$, we try to compute $f(\vec{x})$. If and when we succeed in finding $f(\vec{x})$, we ignore the value and simply say Yes and halt.

To prove (ii) in one direction, suppose that f is an effectively calculable partial function. Here is a semidecision procedure for its graph G: Given $\langle \vec{x}, y \rangle$, we proceed to compute $f(\vec{x})$. If and when we obtain the result, we check to see whether it is y or not. If the result is indeed y, then we say Yes and halt.

Of course, this procedure fails to give an answer if $f(\vec{x}) \uparrow$, which is exactly as it should be, because in this case, $\langle \vec{x}, y \rangle$ is not in the graph.

To prove the other direction of (ii), suppose that we have a semidecision procedure for the graph G. We seek to compute, given $\vec{x}$, the value $f(\vec{x})$, if this is defined. Our plan is to check $\langle \vec{x}, 0 \rangle$, $\langle \vec{x}, 1 \rangle$, ..., for membership in G. But to budget our time sensibly, we use a procedure called "dovetailing." Here is what we do:

1. Spend one minute testing whether $\langle \vec{x}, 0 \rangle \in G$.
2. Spend two minutes testing whether $\langle \vec{x}, 0 \rangle \in G$ and two minutes testing whether $\langle \vec{x}, 1 \rangle \in G$.
3. Similarly, spend three minutes on each of $\langle \vec{x}, 0 \rangle$, $\langle \vec{x}, 1 \rangle$, and $\langle \vec{x}, 2 \rangle$.

And so forth. If and when we discover that, in fact, $\langle \vec{x}, k \rangle \in G$, then we return the value k and halt. Observe that whenever $f(\vec{x}) \downarrow$, then sooner or later the foregoing procedure will correctly determine $f(\vec{x})$ and halt. Of course, if $f(\vec{x}) \uparrow$, then the procedure runs forever.   ⊣

1.1.3   Church's Thesis

Although the concept of effective calculability has here been described in somewhat vague terms, the following section will describe a precise (mathematical) concept of a "computable partial function." In fact, it will describe several equivalent ways of formulating the concept in precise terms. And it will be argued that the mathematical concept of a computable partial function is the *correct* formalization of the informal concept of an effectively calculable partial function. This claim is known as *Church's thesis* or the *Church–Turing thesis*.

Church's thesis, which relates an informal idea to a formal idea, is not itself a mathematical statement capable of being given a proof. But one can look for evidence for or against Church's thesis; it all turns out to be evidence in favor.

One piece of evidence is the absence of counterexamples. That is, any function examined thus far that mathematicians have felt was effectively calculable, has been found to be computable.

Stronger evidence stems from the various attempts that different people made independently, trying to formalize the idea of effective calculability. Alonzo Church used λ-calculus; Alan Turing used an idealized computing agent (later called a Turing machine); Emil Post developed a similar approach. Remarkably, all these attempts turned out to be equivalent, in that they all defined exactly the same class of functions, namely the computable partial functions!

The study of effective calculability originated in the 1930s with work in mathematical logic. As noted previously, the subject is related to the concept of an *acceptable proof*. More recently, the study of effective calculability has formed an essential part of theoretical computer science. A prudent computer scientist would surely want to know that, apart from the difficulties the real world presents, there is a purely theoretical limit to calculability.

Exercises

1. Assume that S is a set of natural numbers containing all but finitely many natural numbers. (That is, S is a *cofinite* subset of $\mathbb{N}$.) Explain why S must be decidable.

2. Assume that A and B are decidable sets of natural numbers. Explain why their intersection $A \cap B$ is also decidable. (Describe an effective procedure for determining whether or not a given number is in $A \cap B$.)

3. Assume that A and B are decidable sets of natural numbers. Explain why their union $A \cup B$ is also decidable.

4. Assume that A and B are semidecidable sets of natural numbers. Explain why their intersection $A \cap B$ is also semidecidable.

5. Assume that A and B are semidecidable sets of natural numbers. Explain why their union $A \cup B$ is also semidecidable.

6. (a) Assume that R is a decidable binary relation on the natural numbers. That is, it is a decidable 2-ary relation. Explain why its domain, $\{x \mid \langle x, y \rangle \in R \text{ for some } y\}$, is a semidecidable set.

   (b) Now suppose that instead of assuming that R is decidable, we assume only that it is semidecidable. Is it still true that its domain must be semidecidable?

7. (a) Assume that f is a one-place total calculable function. Explain why its graph is a decidable binary relation.

   (b) Conversely, show that if the graph of a one-place total function f is decidable, then f must be calculable.

   (c) Now assume that f is a one-place calculable partial function, not necessarily total. Explain why its domain, $\{x \in \mathbb{N} \mid f(x)\downarrow\}$, is semidecidable.

8. Assume that S is a decidable set of natural numbers, and that f is a *total* effectively calculable function on $\mathbb{N}$. Explain why $\{x \mid f(x) \in S\}$ is decidable. (This set is called the *inverse image* of S under f.)

9. Assume that S is a semidecidable set of natural numbers and that f is an effectively calculable partial function on $\mathbb{N}$. Explain why

$$\{x \mid f(x) \downarrow \text{ and } f(x) \in S\}$$

is semidecidable.

10. In the decimal expansion of π, there might be a string of many consecutive 7's. Define the function f so that $f(x) = 1$ if there is a string of x or more consecutive 7's and $f(x) = 0$ otherwise:

$$f(x) = \begin{cases} 1 & \text{if } \pi \text{ has a run of } x \text{ or more 7's} \\ 0 & \text{otherwise.} \end{cases}$$

Explain, without using any special facts about π or any number theory, why f is effectively calculable.

11. Assume that g is a total nonincreasing function on $\mathbb{N}$ (that is, $g(x) \geq g(x+1)$ for all x). Explain why g must be effectively calculable.

12. Assume that f is a total function on the natural numbers and that f is eventually periodic. That is, there exist positive numbers m and p such that for all x greater than m, we have $f(x+p) = f(x)$. Explain why f is effectively calculable.

13. (a) Assume that f is a total effectively calculable function on the natural numbers. Explain why the range of f (that is, the set $\{f(x) \mid x \in \mathbb{N}\}$) is semidecidable.
    (b) Now suppose f is an effectively calculable partial function (not necessarily total). Is it still true that the range must be semidecidable?

14. Assume that f and g are effectively calculable partial functions on $\mathbb{N}$. Explain why the set

$$\{x \mid f(x) = g(x) \text{ and both are defined}\}$$

is semidecidable.

1.2 Formalizations – An Overview

In the preceding section, the concept of effective calculability was described only very informally. Now we want to make those ideas precise (i.e., make them part of mathematics). In fact, several approaches to doing this will be described: idealized computing devices, generative definitions (i.e., the least class containing certain initial functions and closed under certain constructions), programming languages, and definability in formal languages. It is a significant fact that these very different approaches all yield exactly equivalent concepts.

This section gives a general overview of a number of different (but equivalent) ways of formalizing the concept of effective calculability. Later chapters will develop a few of these ways in full detail.

Digression: The 1967 book by Rogers cited in the References demonstrates that the subject of computability can be developed *without* adopting any of these formalizations. And that book was preceded by a 1956 mimeographed preliminary version, which is where I first saw this subject. A few treasured copies of the mimeographed edition still exist.

1.2.1 Turing Machines

In early 1935, Alan Turing was a 22-year-old graduate student at King's College in Cambridge. Under the guidance of Max Newman, he was working on the problem of formalizing the concept of effective calculability. In 1936, he learned of the work of Alonzo Church, at Princeton. Church had also been working on this problem, and in his 1936 article, "An unsolvable problem of elementary number theory," he presented a definite claim that the class of effectively calculable functions should be identified with the class of functions definable in the *lambda calculus*, a formal language for specifying the construction of functions. Church moreover showed that exactly the same class of functions could be characterized in terms of formal derivability from equations.

Turing then promptly completed writing his article, in which he presented a very different approach to characterizing the effectively calculable functions, but one that – as he proved – yielded once again the same class of functions as Church had proposed. With Newman's encouragement, Turing went to Princeton for two years, where he wrote a Ph.D. dissertation under Alonzo Church.

Turing's article remains a very readable introduction to his ideas. How might a diligent clerk carry out a calculation, following instructions? He (or she) might organize the work in a notebook. At any given moment, his attention is focused on a particular page. Following his instructions, he might alter that page, and then he might turn to another page. And the notebook is large enough (or the supply of fresh paper is ample enough) that he never comes to the last page.

The alphabet of symbols available to the clerk must be finite; if there were infinitely many symbols, then there would be two that were arbitrarily similar and so might be confused. We can then without loss of generality regard what can be written on one page of notebook as a single symbol. And we can envision the notebook pages as being placed side by side, forming a paper tape, consisting of squares, each square being either blank or printed with a symbol. (For uniformity, we can think of a blank square as containing the "blank" symbol *B*.) At each stage of his work, the clerk – or the mechanical machine – can alter the square under examination, can turn attention to the next square or the previous one, and can look to the instructions to see what part of them to follow next. Turing described the latter part as a "change of state of mind."

Turing wrote, "We may now construct a machine to do the work." Such a machine is, of course, now called a *Turing machine*, a phrase first used by Church in his review of Turing's article in *The Journal of Symbolic Logic*. The machine has a potentially infinite tape, marked into squares. Initially the given input numeral or word is written on the tape, which is otherwise blank. The machine is capable of being in any one of finitely many "states" (the phrase "of mind" being inappropriate for a machine).

At each step of calculation, depending on its state at the time, the machine can change the symbol in the square under examination at that time, and can turn its attention to the square to the left or to the right, and can then change its state to another state. (The tape stretches endlessly in both directions.)

The program for this Turing machine can be given by a table. Where the possible states of the machine are $q_1, \ldots, q_r$, each line of the table is a quintuple $\langle q_i, S_j, S_k, D, q_m \rangle$ which is to be interpreted as directing that whenever the machine is in state q_i and the square under examination contains the symbol S_j, then that symbol should be altered to S_k and the machine should shift its attention to the square to the left (if $D = L$) or to the right (if $D = R$), and should change its state to q_m. Possibly S_j is the "blank" symbol B, meaning the square under examination is blank; possibly S_k is B, meaning that whatever is in the square is to be erased. For the program to be unambiguous, it should have no two different quintuples with the same first two components. (By relaxing this requirement regarding absence of ambiguity, we obtain the concept of a *nondeterministic* Turing machine, which will be useful later, in the discussion of feasible computability.) One of the states, say, q_1, is designated as the initial state – the state in which the machine begins its calculation. If we start the machine running in this state, and examining the first square of its input, it might (or might not), after some number of steps, reach a state and a symbol for which its table lacks a quintuple having that state and symbol for its first two components. At that point the machine *halts*, and we can look at the tape (starting with the square which was then under examination) to see what the output numeral or word is.

Now suppose that Σ is a finite alphabet (the blank B does not count as a member of Σ). Let Σ^* be the set of all words over this alphabet (that is, Σ^* is the set of all strings, including the empty string, consisting of members of Σ). Suppose that f is a k-place partial function from Σ^* into Σ^*. We will say that f is *Turing computable* if there exists a Turing machine $\mathcal{M}$ that, when started in its initial state scanning the first symbol of a k-tuple $\vec{w}$ of words (written on the tape, with a blank square between words, and with the rest of the tape blank), behaves as follows:

- If $f(\vec{w}) \downarrow$ (i.e., if $\vec{w} \in \mathrm{dom} f$) then $\mathcal{M}$ eventually halts, and at that time, it is scanning the leftmost symbol of the word $f(\vec{w})$ (which is followed by a blank square).
- If $f(\vec{w}) \uparrow$ (i.e., if $\vec{w} \notin \mathrm{dom} f$) then $\mathcal{M}$ never halts.

Example: Take a two-letter alphabet $\Sigma = \{a, b\}$. Let $\mathcal{M}$ be the Turing machine given by the following set of six quintuples, where q_1 is designated as the initial state:

$$\langle q_1, a, a, R, q_1 \rangle$$
$$\langle q_1, b, b, R, q_1 \rangle$$
$$\langle q_1, B, a, L, q_2 \rangle$$
$$\langle q_2, a, a, L, q_2 \rangle$$
$$\langle q_2, b, b, R, q_2 \rangle$$
$$\langle q_2, B, B, R, q_3 \rangle.$$

Suppose we start this machine in state q_1, scanning the first letter of a word w. The machines move (in state q_1) to the right end of w, where it appends the letter a. Then it moves (in state q_2) back to the left end of the word, where it halts (in state q_3). Thus, $\mathcal{M}$ computes the total function $f(w) = wa$.

We need to adopt special conventions for handling the empty word λ, which occupies zero squares. This can be done in different ways; the following is the way chosen. If the machine halts scanning a blank square, then the output word is λ. For a one-place function f, to compute $f(\lambda)$, we simply start with a blank tape. For a two-place function g, to compute $g(w, \lambda)$, we start with only the word w, scanning the first symbol of w. And to compute $g(\lambda, w)$, we also start with only the word w on the tape, but scanning the blank square just to the left of w. And in general, to give a k-place function the input $\vec{w} = \langle u_1, \ldots, u_k \rangle$ consisting of k words of lengths $n_1, \ldots, n_k$, we start the machine scanning the first square of the input configuration of length $n_1 + \cdots + n_k + k - 1$

$$(n_1 \text{ symbols from } u_1)B(n_2 \text{ symbols from } u_2)B \cdots B(n_k \text{ symbols from } u_k)$$

with the rest of the tape blank. Here any n_i can be zero; in the extreme case, they can all be zero.

An obvious drawback of these conventions is that there is no difference between the pair $\langle u, v \rangle$ and the triple $\langle u, v, \lambda \rangle$. Other conventions avoid this drawback, at the cost of introducing their own idiosyncrasies.

The definition of Turing computability can be readily adapted to apply to k-place partial functions on $\mathbb{N}$. The simplest way to do this is to use base-1 numerals. We take a one-letter alphabet $\Sigma = \{|\}$ whose one letter is the tally mark $|$. Or to be more conventional, let $\Sigma = \{1\}$, using the symbol 1 in place of the tally mark. Then the input configuration for the triple $\langle 3, 0, 4 \rangle$ is

111BB1111.

Then *Church's thesis*, also called – particularly in the context of Turing machines – the *Church–Turing thesis*, is the claim that this concept of Turing computability is the correct formalization of the informal concept of effective calculability. Certainly the definition reflects the ideas of following predetermined instructions, without limitation of the amount of time that might be required. (The name "Church–Turing thesis" obscures the fact that Church and Turing followed very different paths in reaching equivalent conclusions.)

Church's thesis has by now achieved universal acceptance. Kurt Gödel, writing in 1964 about the concept of a "formal system" in logic, involving the idea that the set of correct deductions must be a decidable set, said that "due to A. M. Turing's work, a precise and unquestionably adequate definition of the general concept of formal system can now be given." And others agree.

The robustness of the concept of Turing computability is evidenced by the fact that it is insensitive to certain modifications to the definition of a Turing machine. For example, we can impose limitations on the size of the alphabet, or we can insist that the machine never moves to the left of its initial starting point. None of this will affect that class of Turing computable partial functions.

Turing developed these ideas before the introduction of modern digital computers. After World War II, Turing played an active role in the development of early computers, and in the emerging field of artificial intelligence. (During the war, he worked on deciphering the German battlefield code Enigma, militarily important work, which remained classified until after Turing's death.) One can speculate as to whether Turing might have formulated his ideas somewhat differently, if his work had come after the introduction of digital computers.

Digression: There is an interesting example, here, that goes by the name[1] of "the busy beaver problem."

Suppose we want a Turing machine, starting on a blank tape, to write as many 1's as it can, and then stop. With a limited number of states, how many 1's can we get?

To make matters more precise, take Turing machines with the alphabet $\{1\}$ (so the only symbols are B and 1). We will allow such machines to have n states, plus a halting state (that can occur as the last member of a quintuple, but not as the first member). For each n, there are only finitely many essentially different such Turing machines. Some of them, started on a blank tape, might not halt. For example, the one-state machine

$$\langle q_1, B, 1, R, q_1 \rangle$$

keeps writing forever without halting. But among those that do halt, we seek the ones that write a lot of 1's.

Define $\sigma(n)$ to be the largest number of 1's that can be written by an n-state Turing machine as described here before it halts. For example, $\sigma(1) = 1$, because the one-state machine

$$\langle q_1, B, 1, R, q_H \rangle$$

(the halting state q_H doesn't count) writes one 1, and none of the other one-state machines do any better. (There are not so very many one-state machines, and one can examine all of them in a reasonable length of time.) Let's agree that $\sigma(0) = 0$. Then σ is a total function. It is also nondecreasing because having an extra state to work with is never a handicap. Despite the fact that $\sigma(n)$ is merely the largest member of a certain finite set, there is no algorithm that lets us, in general, evaluate it.

Example: Here is a two-state candidate:

$$\langle q_1, B, 1, R, q_2 \rangle$$
$$\langle q_1, 1, 1, L, q_2 \rangle$$
$$\langle q_2, B, 1, L, q_1 \rangle$$
$$\langle q_2, 1, 1, R, q_H \rangle$$

Started on a blank tape, this machine writes four consecutive 1's, and then halts (after six steps), scanning the third 1. You are invited to verify this by running the machine. We conclude the $\sigma(2) \geq 4$.

---

[1] This name has given translators much difficulty.

Rado's theorem (1962): *The function σ is not Turing computable. Moreover, for any Turing computable total function f, we have $f(x) < \sigma(x)$ for all sufficiently large x. That is, σ eventually dominates any Turing computable total function.*

Proof. Assume we are given some Turing computable total f. We must show that σ eventually dominates it. Define (for reasons that may initially appear mysterious) the function g:

$$g(x) = \max(f(2x), f(2x+1)) + 1.$$

Then g is total and one can show that it is Turing computable. So there is some Turing machine $\mathcal{M}$ with, say, k states that computes it, using the alphabet $\{1\}$ and base-1 notation. For each x, let $\mathcal{N}_x$ be the $(x + k)$-state Turing machine that first writes x 1's on the tape, and then imitates $\mathcal{M}$. (The x states let us write x 1's on the tape in a straightforward way, and then there are the k states in $\mathcal{M}$.)

Then $\mathcal{N}_x$, when started on a blank tape, writes $g(x)$ 1's on the tape and halts. So $g(x) \le \sigma(x + k)$, by the definition of σ. Thus, we have

$$f(2x), f(2x+1) < g(x) \le \sigma(x+k),$$

and if $x \ge k$, then

$$\sigma(x+k) \le \sigma(2x) \le \sigma(2x+1).$$

Putting these two lines together, we see that $f < \sigma$ from $2k$ on.   ⊣

So σ grows faster – eventually – than any Turing computable total function. How fast does it grow? Among the smaller numbers, $\sigma(2) = 4$. (The preceding example shows that $\sigma(2) \ge 4$. The other inequality is not entirely trivial because there are thousands of two-state machines.) It has also been shown that $\sigma(3) = 6$ and $\sigma(4) = 13$. From here on, only lower bounds are known. In 1984, it was found that $\sigma(5)$ is at least 1915. In 1990, this was raised to 4098. And $\sigma(6) > 3.1 \times 10^{10\,566}$. And $\sigma(7)$ must be astronomical. These lower bounds are established by using ingeniously convoluted coding to make small Turing machines that write that many 1's and then halt.

Proving further *upper* bounds would be difficult. In fact, one can show, under some reasonable assumptions, that upper bounds on $\sigma(n)$ are provable for only finitely many n's.

If we could solve the halting problem, we would then have the following method for computing $\sigma(n)$:

- List all the n-state machines.
- Discard those that never halt.
- Run those that do halt.
- Select the highest score.

It is the second step in this method that gives us trouble. (New information on Rado's σ function continues to be discovered. Recent news can be obtained from the Web page maintained by Heiner Marxen, http://www.drb.insel.de/~heiner/BB.)

1.2.2   Primitive Recursiveness and Search

For a second formalization of the calculability concept, we will define a certain class of partial functions on $\mathbb{N}$ as the smallest class that contains certain initial functions and is closed under certain constructions.

For the initial functions, we take the following very simple total functions:

- The *zero* functions, that is, the constant functions f defined by the equation:

$$f(x_1, \ldots, x_k) = 0.$$

  There is one such function for each k.
- The *successor* function S, defined by the equation:

$$S(x) = x + 1.$$

- The *projection* functions I_n^k from k-dimensions onto the nth coordinate,

$$I_n^k(x_1, \ldots, x_k) = x_n,$$

  where $1 \leq n \leq k$.

We want to form the closure of the class of initial functions under three constructions: composition, primitive recursion, and search.

A k-place function h is said to be obtained by *composition* from the n-place function f and the k-place functions $g_1, \ldots, g_n$ if the equation

$$h(\vec{x}) = f(g_1(\vec{x}), \ldots, g_n(\vec{x}))$$

holds for all $\vec{x}$. In the case of partial functions, it is to be understood here that $h(\vec{x})$ is undefined unless $g_1(\vec{x}), \ldots, g_n(\vec{x})$ are all defined and $\langle g_1(\vec{x}), \ldots, g_n(\vec{x}) \rangle$ belongs to the domain of f.

A $(k + 1)$-place function h is said to be obtained by *primitive recursion* from the k-place function f and the $(k + 2)$-place function g (where $k > 0$) if the pair of equations

$$h(\vec{x}, 0) = f(\vec{x})$$
$$h(\vec{x}, y + 1) = g(h(\vec{x}, y), \vec{x}, y)$$

holds for all $\vec{x}$ and y.

Again, in the case of partial functions, it is to be understood that $h(\vec{x}, y + 1)$ is undefined unless $h(\vec{x}, y)$ is defined and $\langle h(\vec{x}, y), \vec{x}, y \rangle$ is in the domain of g.

Observe that in this situation, knowing the two functions f and g completely determines the function h. More formally, if h_1 and h_2 are both obtained by primitive recursion from f and g, then for each $\vec{x}$, we can show by induction on y that $h_1(\vec{x}, y) = h_2(\vec{x}, y)$.

For the $k = 0$ case, the one-place function h is obtained by primitive recursion from the two-place function g by using the number m if the pair of equations

$$h(0) = m$$
$$h(y + 1) = g(h(y), y)$$

holds for all y.

Postponing the matter of search, we define a function to be *primitive recursive* if it can be built up from zero, successor, and projection functions by the use of composition and primitive recursion. (See the beginning of Chapter 2 for some examples.) In other words, the class of primitive recursive functions is the smallest class that includes our initial functions and is closed under composition and primitive recursion. (Here saying that a class C is "closed" under composition and primitive recursion means that whenever a function f is obtained by composition from functions in C or is obtained by primitive recursion from functions in C, then f itself also belongs to C.)

Clearly all the primitive recursive functions are total. This is because the initial functions are all total, the composition of total functions is total, and a function obtained by primitive recursion from total functions will be total.

We say that a k-ary relation R on $\mathbb{N}$ is primitive recursive if its characteristic function is primitive recursive.

One can then show that a great many of the common functions on $\mathbb{N}$ are primitive recursive: addition, multiplication, ..., the function whose value at m is the $(m+1)$st prime, ... Chapter 2 will carry out the project of showing that many functions are primitive recursive.

On the one hand, it seems clear that every primitive recursive function should be regarded as being effectively calculable. (The initial functions are pretty easy. Composition presents no big hurdles. Whenever h is obtained by primitive recursion from effectively calculable f and g, then we see how we could effectively find $h(\vec{x}, 99)$, by first finding $h(\vec{x}, 0)$ and then working our way up.) On the other hand, the class of primitive recursive functions cannot possibly comprehend all total calculable functions because we can "diagonalize out" of the class. That is, by suitably indexing the "family tree" of the primitive recursive functions, we can make a list $f_0, f_1, f_2, \ldots$ of all the one-place primitive recursive functions. Then consider the diagonal function $d(x) = f_x(x) + 1$. Then d cannot be primitive recursive; it differs from each f_x at x. Nonetheless, if we made our list very tidily, the function d will be effectively calculable. The conclusion is the class of primitive recursive functions is an extensive but proper subset of the total calculable functions.

Next, we say that a k-place function h is obtained from the $(k+1)$-place function g by *search*, and we write

$$h(\vec{x}) = \mu\, y[g(\vec{x}, y) = 0]$$

if for each $\vec{x}$, the value $h(\vec{x})$ either is the number y such that $g(\vec{x}, y) = 0$ and $g(\vec{x}, s)$ is defined and is nonzero for every $s < y$, if such a number t exists, or else is undefined, if no such number t exists. The idea behind this "μ-operator" is the idea of searching for the least number y that is the solution to an equation, by testing successively $y = 0, 1, \ldots$.

We obtain the *general recursive* functions by adding search to our closure methods. That is, a partial function is general recursive if it can be built up from the initial zero, successor, and projection functions, by use of composition, primitive recursion, and search (i.e, the μ-operator).

The class of general recursive partial functions on $\mathbb{N}$ is (as Turing proved) exactly the same as the class of Turing computable partial functions. This is a rather striking

result, in light of the very different ways in which the two definitions were formulated. Turing machines would seem, at first glance, to have little to do with primitive
recursion and search. And yet, we get exactly the same partial functions from the two
approaches. And Church's thesis, therefore, has the equivalent formulation that the
concept of a general recursive function is the correct formalization of the informal
concept of effective calculability.

What if we try to "diagonalize out" of the class of general recursive functions, as
we did for the primitive recursive functions? As will be argued later, we can again
make a tidy list $\varphi_0, \varphi_1, \varphi_2, \ldots$ of all the one-place general recursive partial functions.
And we can define the diagonal function $d(x) = \varphi_x(x) + 1$. But in this equation, $d(x)$ is
undefined unless $\varphi_x(x)$ is defined. The diagonal function d is indeed among the general
recursive partial functions, and hence is φ_k for some k, but $d(k)$ must be undefined.
No contradiction results.

The class of primitive recursive functions was defined by Gödel, in his 1931 article
on the incompleteness theorems in logic. Of course, the idea of defining functions on
$\mathbb{N}$ by recursion is much older, and reflects the idea that the natural numbers are built
up from the number 0 by repeated application of the successor function. (Dedekind
wrote about this topic.) The theory of the general recursive functions was worked out
primarily by Stephen Kleene, a student of Church.

The use of the word "recursive" in the context of the primitive recursive functions
is entirely reasonable. Gödel, writing in German, had used simply "rekursiv" for the
primitive recursive functions. (It was Rózsa Péter who introduced the term "primitive
recursive.") But the class of general recursive functions has – as this section shows –
several other characterizations in which *recursion* (i.e., defining a function in terms of
its other values, or using routines that call themselves) plays no obvious role.

This leads to the question: What to call this class of functions? Having two names
("Turing computable" and "general recursive") is an embarrassment of riches, and the
situation will only grow worse. Historically, the name "partial recursive functions"
won out. And relations on $\mathbb{N}$ were said to be *recursive* if their characteristic functions
belonged to the class. The study of such functions was for years called "recursive
function theory," and then "recursion theory." But this was more a matter of histori-
cal accident than a matter of reasoned choice. Nonetheless, the terminology became
standard.

But now an effort is being made to change what had been the standard terminology. Accordingly, this book, *Computability Theory*, speaks of *computable* partial
functions. And we will call a relation *computable* if its characteristic function is a
computable function. Thus, the concept of a computable relation corresponds to the
informal notion of a decidable relation. (The manuscript for this book has, however,
been prepared with TeX macros that would facilitate a rapid change in terminology.)

In any case, there is definitely a need to have separate adjectives for the informal
concept (here "calculable" is used for functions, and "decidable" for relations) and the
formally defined concept (here "computable").

1.2.3   Loop and While Programs

The idea behind the concept of effective calculable functions is that one should be
able to give explicit instructions – a program – for calculating such a function. What

programming language would be adequate here? Actually, any of the commonly used programming languages would suffice, if freed from certain practical limitations, such as the size of the number denoted by a variable. We give here a simple programming language with the property that the programmable functions are exactly the computable partial functions on $\mathbb{N}$.

The variables of the language are $X_0, X_1, X_2, \ldots$. Although there are infinitely many variables in the language, any one program, being a finite string of commands, can have only a finite number of these variables. If we want the language to consist of words over a finite alphabet, we can replace X_3, say, by X'''.

In running a program, each variable in the program gets assigned a natural number. There is no limit on how large this number can be. Initially, some of the variables will contain the input to the function; the language has no "input" commands. Similarly, the language has no "output" commands; when (and if) the program halts, the value of X_0 is to be the function value.

The commands of the language come in five kinds:

1. $X_n \leftarrow 0$. This is the *clear* command; its effect is to assign the value 0 to X_n.
2. $X_n \leftarrow X_n + 1$. This is the *increment* command; its effect is to increase the value assigned to X_n by one.
3. $X_n \leftarrow X_m$. This is the *copy* command; its effect is just what the name suggests; in particular, it leaves the value of X_m unchanged.
4. loop X_n and endloop X_n. These are the *loop* commands, and they must be used in pairs. That is, if $\mathcal{P}$ is a program – a syntactically correct string of commands – then so is the string:

   > loop X_n
   > $\mathcal{P}$
   > endloop X_n

   What this program means is that $\mathcal{P}$ is to be executed a certain number k of times. And that number k is the *initial* value of X_n, the value assigned to X_n before we start executing $\mathcal{P}$. Possibly $\mathcal{P}$ will change the value of X_n; this has no effect at all on k. If $k = 0$, then this string does nothing.
5. while $X_n \neq 0$ and endwhile $X_n \neq 0$. These are the *while* commands; again, they must be used in pairs, like the loop commands. But there is a difference. The program

   > while $X_n \neq 0$
   > $\mathcal{P}$
   > endwhile $X_n \neq 0$

   also executes the program $\mathcal{P}$ some number k of times. But now k is *not* determined in advance; it matters very much how $\mathcal{P}$ changes the value of X_n. The number k is the least number (if any) such that executing $\mathcal{P}$ that many times causes X_n to be assigned the value 0. The program will run forever if there is no such k.

And those are the only commands. A *while* program is a sequence of commands, subject only to the requirement that the loop and while commands are used in pairs, as illustrated. Clearly, this programming language is simple enough to be simulated by any of the common programming languages if we ignore overflow problems.

A *loop* program is a while program with no while commands; that is, it has only clear, increment, copy, and loop commands. Note the important property: A loop program *always halts*, no matter what. But it is easy to make a while program that never halts.

We say that a k-place partial function f on $\mathbb{N}$ is *while-computable* if there exists a while program $\mathcal{P}$ that, whenever started with a k-tuple $\vec{x}$ assigned to the variables $X_1, \ldots, X_k$ and 0 assigned to the other variables, behaves as follows:

- If $f(\vec{x})$ is defined, then the program eventually halts, with X_0 assigned the value $f(\vec{x})$.
- If $f(\vec{x})$ is undefined, then the program never halts.

The *loop-computable* functions are defined in the analogous way. But there is the difference that any loop-computable function is total.

Theorem:

(a) *A function on $\mathbb{N}$ is loop-computable if and only if it is primitive recursive.*

(b) *A partial function on $\mathbb{N}$ is while-computable if and only if it is general recursive.*

The proof in one direction, to show that every primitive recursive function is loop-computable, involves a series of programming exercises. The proof in the other direction involves coding the status of a program $\mathcal{P}$ on input $\vec{x}$ after t steps, and showing that there are primitive recursive functions enabling us to determine the status after $t + 1$ steps, and the terminal status.

Because the class of general recursive partial functions coincides with the class of Turing computable partial functions, we can conclude from the above theorem that while-computability coincides with Turing computability.

1.2.4 Register Machines

Here is another programming language. On the one hand, it is extremely simple – even simpler than the language for loop-while programs. On the other hand, the language is "unstructured"; it incorporates (in effect) go-to commands. This formalization was presented by Shepherdson and Sturgis in a 1963 article.

A *register machine* is to be thought of as a computing device with a finite number of "registers," numbered $0, 1, 2, \ldots, K$. Each register is capable of storing a natural number of any magnitude – there is no limit to the size of this number. The operation of the machine is determined by a *program*. A program is a finite sequence of *instructions*, drawn from the following list:

- "Increment r," I r (where $0 \leq r \leq K$): The effect of this instruction is to increase the contents of register r by 1. The machine then proceeds to the next instruction in the program (if any).
- "Decrement r," D r (where $0 \leq r \leq K$): The effect of this instruction depends on the contents of register r. If that number is nonzero, it is decreased by 1, and the machine proceeds *not* to the next instruction, but to the following one. But if the number in register r is zero, the machine simply proceeds to the next instruction. In summary, the machine tries to decrement register r, and if it is successful, then it skips one instruction.

- "Jump q," J q (where q is an integer – positive, negative, or zero): All registers are left unchanged. The machine takes as its next instruction the qth instruction following this one in the program (if $q \geq 0$), or the $|q|$th instruction preceding this one (if $q < 0$). The machine halts if there is no such instruction in the program. An instruction of J 0 results in a loop, with the machine executing this one instruction over and over again.

And that is all. The language has only these three types of instructions. (Strictly speaking, in these instructions, r and q are numerals, not numbers. That is, an instruction should be a sequence of *symbols*. If we use base-10 numerals, then the alphabet is $\{I, D, J, 0, 1, 2, 3, 4, 5, 6, 7, 8, 9, -\}$. An instruction is a correctly formed word over this alphabet.)

Examples:

1. CLEAR 7: a program to clear register 7.

D	7	Try to decrement 7.
J	2	Go back and repeat.
J	-2	Halt.

2. MOVE from r to s: a program to move a number from register r to register s (where $r \neq s$).

CLEAR s.		Use the program of the first example.
D	r	Take 1 from r.
J	3	Halt when zero.
I	s	Add 1 to s.
J	-3	Repeat.

    This program has seven instructions altogether. It leaves a zero in register r.

3. ADD 1 to 2 and 3: a program to add register 1 to registers 2 and 3.

D	1
J	4
I	2
I	3
J	-4

    This program leaves a zero in register 1. It is clear how to adapt the program to add register 1 to more (or fewer) than two registers.

4. COPY from r to s (using t): a program to copy a number from register r to register s (leaving register r unchanged). We combine the previous examples.

CLEAR s.	Use the first example.
MOVE from r to t.	Use the second example.
ADD t to r and s.	Use the third example.

    This program has 15 instructions. It uses a third register, register t. At the end, the contents for register r are restored. But during execution, register r must be cleared; this is the only way of determining its contents. (It is assumed here that r, s, and t are distinct.)

5. (Addition) Say that x and y are in registers 1 and 2. We want $x + y$ in register 0, and we want to leave x and y still in registers 1 and 2 at the end.

	Register contents			
CLEAR 0.	0	x	y	
MOVE from 1 to 3.	0	0	y	x
ADD 3 to 1 and 0.	x	x	y	0
MOVE from 2 to 3.	x	x	0	y
ADD 3 to 2 and 0.	$x + y$	x	y	0

This program has 27 instructions as it is written, but three of them are unnecessary. (In the fourth line, we begin by clearing register 3, which is already clear.)

Now suppose f is an n-place partial function on $\mathbb{N}$. Possibly, there will be a program $\mathcal{P}$ such that if we start a register machine (having all the registers to which $\mathcal{P}$ refers) with $x_1, \ldots, x_n$ in registers $1, \ldots, n$ and 0 in the other registers, and we apply program $\mathcal{P}$, then the following conditions hold:

- If $f(x_1, \ldots, x_n)$ is defined, then the computation eventually terminates with $f(x_1, \ldots, x_n)$ in register 0. Furthermore, the computation terminates by seeking a $(p+1)$st instruction, where p is the length of $\mathcal{P}$.
- If $f(x_1, \ldots, x_n)$ is undefined, then the computation never terminates.

If there is such a program $\mathcal{P}$, we say that $\mathcal{P}$ *computes f.*

Which functions are computable by register-machine programs? The language is so simple – it appears to be a toy language – that one's first impression might be that only very simple functions are computable. This impression is misleading.

Theorem: *Let f be a partial function. Then, there is a register-machine program that computes f if and only if f is a general recursive partial function.*

Thus by using register machines, we arrive at exactly the class of general recursive partial functions, a class we originally defined in terms of primitive recursion and search.

1.2.5 Definability in Formal Languages

We will briefly sketch several other ways in which the concept of effective calculability might be formalized. Details will be left to the imagination.

In 1936, in his article in which he presented what is now known as Church's thesis, Alonzo Church utilized a formal system, the λ-*calculus*. Church had developed this system as part of his study of the foundations of logic. In particular, for each natural number, n, there is a formula $\bar{n}$ of the system denoting n, that is, a numeral for n. More importantly, formulas could be used to represent the construction of functions. He defined a two-place function F to be λ-*definable* if there existed a formula $\mathbf{F}$ of the lambda calculus such that whenever $F(m, n) = r$, then the formula $\{\mathbf{F}\}(\bar{m}, \bar{n})$ was convertible, following the rules of the system, to the formula $\bar{r}$, and only then. An analogous definition applied to k-place functions.

Church's student Stephen Kleene showed that a function was λ-definable if and only if it was general recursive. (Church and his student J. B. Rosser also were involved in the development of this result.) Church wrote in his article, "The fact ... that two such widely different and (in the opinion of the author) equally natural definitions of effective calculability turn out to be equivalent adds to the strength of reasons ... for believing that they constitute as general a characterization of this notion as is consistent with the usual intuitive understanding of it."

Earlier, in 1934, Kurt Gödel, in lectures at Princeton, formulated a concept now referred to as Gödel–Herbrand computability. He did not, however, at the time propose the concept as a formalization of the concept of effective calculability. The concept involved a formal calculus of equations between terms built up from variables and function symbols. The calculus permitted the passage from an equation $A = B$ to another equation obtained by substituting for a part C of A or B another term D where the equation $C = D$ had been derived. If a set $\mathcal{E}$ of equations allowed the derivation, in a suitable sense, of exactly the right values for a function f on $\mathbb{N}$, then $\mathcal{E}$ was said to be a set of *recursion equations* for f. Once again, it turned out that a set of recursion equations existed for f if and only if f was a general recursive function.

A rather different approach to characterizing the effectively calculable functions involves definability by expressions in symbolic logic. A formal language for the arithmetic of natural numbers might have variables and a numeral for each natural number, and symbols for the equality relation and for the operations of addition and multiplication, at least. Moreover, the language should be able to handle the basic logical connectives such as "and," "or," and "not." Finally, it should include the "quantifier" expressions $\forall v$ and $\exists v$ meaning "for all natural numbers v" and "for some natural number v," respectively.

For example,

$$\exists s(u_1 + s = u_2)$$

might be an expression in the formal language, asserting a property of u_1 and u_2. The expression is true (in $\mathbb{N}$ with its usual operations) when u_1 is assigned 4 and u_2 is assigned 9 (take $s = 5$). But it is false when u_1 is assigned 9 and u_2 is assigned 4. More generally, we can say that the expression *defines* (in $\mathbb{N}$ with its usual operations) the binary relation "$\leq$" on $\mathbb{N}$.

For another example,

$$v \neq 0 \text{ and } \forall x \forall y [\exists s(v + s = x) \text{ or } \exists t(v + t = y) \text{ or } v \neq x \cdot y]$$

might be an expression in the formal language, asserting a property of v. The expression is false (in $\mathbb{N}$ with its usual operation) when v is assigned the number 6 (try $x = 2$ and $y = 3$). But the expression is true when v is assigned 7. More generally, the expression is true when v is assigned a prime number, and only then. We can say that this expression *defines* the set of prime numbers (in $\mathbb{N}$ with its usual operations).

Say that a k-place partial function f on $\mathbb{N}$ is Σ_1-*definable* if the graph of f (that is, the $(k + 1)$-ary relation $\{\langle \vec{x}, y \rangle \mid f(\vec{x}) = y\}$) can be defined in $\mathbb{N}$ and with the operations of addition, multiplication, and exponentiation, by an expression of the following form:

$$\exists v_1 \exists v_2 \cdots \exists v_n (\text{expression without quantifiers})$$

Then the class of Σ_1-definable partial functions coincides exactly with the class of partial functions given by the other formalizations of calculability described here. Moreover, Yuri Matiyasevich showed in 1970 that the operation of exponentiation was not needed here.

Finally, say that a k-place partial function f on $\mathbb{N}$ is *representable* if there exists some finitely axiomatizable theory T in a language having a suitable numeral $\bar{n}$ for each natural number n, and there exists a formula φ of that language such that (for any natural numbers) $f(x_1, \ldots, x_k) = y$ if and only if $\varphi(\bar{x}_1, \ldots, \bar{x}_k, \bar{y})$ is a sentence deducible in the theory T. Then, once again, the class of representable partial functions coincides exactly with the class of partial functions given by the other formalizations of calculability described here.

1.2.6 Church's Thesis Revisited

In summary, for a k-place partial function f, the following conditions are equivalent:

- The function f is a Turing-computable partial function.
- The function f is a general recursive partial function.
- The partial function f is while-computable.
- The partial function f is computed by some register-machine program.
- The partial function f is λ-definable.
- The partial function f is Σ_1-definable (over the natural numbers with addition, multiplication, and exponentiation).
- The partial function f is representable (in some finitely axiomatizable theory).

The equivalence of these conditions is surely a remarkable fact! Moreover, it is evidence that the conditions characterize some natural and significant property. Church's thesis is the claim that the conditions in fact capture the informal concept of an effectively calculable function.

Definition: A k-place partial function f on the natural numbers is said to be a *computable partial function* if the foregoing conditions hold.

Then Church's thesis is the claim that this definition is the one we want.

The situation is somewhat analogous to one in calculus. An intuitively continuous function (defined on an interval) is one whose graph can be drawn without lifting the pencil off the paper. But to prove theorems, some formalized counterpart of this concept is needed. And so one gives the usual definition of ε-δ-continuity. Then it is fair to ask whether the precise concept of ε-δ-continuity is an accurate formalization of intuitive continuity. If anything, the class of ε-δ-continuous functions is too *broad*. It includes nowhere differentiable functions, whose graphs cannot be drawn without lifting the pencil – there is no way to impart a velocity vector to the pencil. But accurate or not, the class of ε-δ-continuous functions has been found to be a natural and important class in mathematical analysis.

Very much the same situation occurs with computability. It is fair to ask whether the precise concept of a computable partial function is an accurate formalization of the informal concept of an effectively calculable function. Again, the precisely defined class appears to be, if anything, too broad, because it includes functions requiring, for large inputs, absurd amounts of computing time. Computability corresponds to effective calculability in an idealized world, where length of computation and amount of memory space are disregarded. But in any event, the class of computable partial functions has been found to be a natural and important class.

Exercises

15. Give a loop program to compute the following function:

$$f(x, y, z) = \begin{cases} y & \text{if } x = 0 \\ z & \text{if } x \neq 0. \end{cases}$$

16. Let $x \dotminus y = \max(x - y, 0)$, the result of subtracting y from x, but with a "floor" of 0. Give a loop program that computes the two-place function $x \dotminus y$.

17. Give a loop program that when started with all variables assigned 0, halts with X_0 assigned some number greater than 1000.

18. **(a)** Give a register-machine program that computes the subtraction function, $x \dotminus y = \max(x - y, 0)$, as in Exercise 16.

   (b) Give a register-machine program that computes the subtraction partial function:

$$f(x, y) = \begin{cases} x - y & \text{if } x \geq y \\ \uparrow & \text{if } x < y. \end{cases}$$

19. Give a register-machine program that computes the multiplication function, $x \cdot y$.

20. Give a register-machine program that computes the function $\max(x, y)$.

21. Give a register-machine program that computes the parity function:

$$f(x) = \begin{cases} 1 & \text{if } x \text{ is odd} \\ 0 & \text{if } x \text{ is even.} \end{cases}$$

2 General Recursive Functions

In the preceding chapter, we saw an overview of several possible formalizations of the concept of effective calculability. In this chapter, we focus on *one* of those: primitive recursiveness and search, which give us the class of general recursive partial functions. In particular, we develop tools for showing that certain functions are in this class. These tools will be used in Chapter 3, where we study computability by register-machine programs.

2.1 Primitive Recursive Functions

The primitive recursive functions have been defined in the preceding chapter as the functions on $\mathbb{N}$ that can be built up from zero functions

$$f(x_1, \ldots, x_k) = 0,$$

the successor function

$$S(x) = x + 1,$$

and the projection functions

$$I_n^k(x_1, \ldots, x_k) = x_n$$

by using (zero or more times) composition

$$h(\vec{x}) = f(g_1(\vec{x}), \ldots, g_n(\vec{x}))$$

and primitive recursion

$$h(\vec{x}, 0) = f(\vec{x})$$
$$h(\vec{x}, y + 1) = g(h(\vec{x}, y), \vec{x}, y),$$

where $\vec{x}$ can be empty:

$$h(0) = m$$
$$h(y + 1) = g(h(y), y).$$

Example: Suppose we are given the number $m = 1$ and the function $g(w, y) = w \cdot (y + 1)$. Then the function h obtained by primitive recursion from g by using m is the

Computability Theory

function given by the pair of equations

$$h(0) = m = 1$$
$$h(y+1) = g(h(y), y) = h(y) \cdot (y+1).$$

Using this pair of equations, we can proceed to calculate the values of the function h:

$$h(0) = m = 1$$
$$h(1) = g(h(0), 0) = g(1, 0) = 1$$
$$h(2) = g(h(1), 1) = g(1, 1) = 2$$
$$h(3) = g(h(2), 2) = g(2, 2) = 6$$
$$h(4) = g(h(3), 3) = g(6, 3) = 24$$

And so forth. In order to calculate $h(4)$, we first need to know $h(3)$, and to find that we need $h(2)$, and so on. The function h in this example is, of course, better known as the factorial function, $h(x) = x!$.

It should be pretty clear that given any number m and any two-place function g, *there exists* a unique function h obtained by primitive recursion from g by using m. It is the function h that we calculate as in the preceding example. Similarly, given a k-place function f and a $(k + 2)$-place function g, there exists a unique $(k + 1)$-place function h that is obtained by primitive recursion from f and g. That is, h is the function given by the pair of equations

$$h(\vec{x}, 0) = f(\vec{x})$$
$$h(\vec{x}, y+1) = g(h(\vec{x}, y), \vec{x}, y).$$

Moreover, if f and g are total functions, then h will also be total.

Example: Consider the addition function $h(x, y) = x + y$. For any fixed x, its value at $y + 1$ (i.e., $x + y + 1$) is obtainable from its value at y (i.e., $x + y$) by the simple step of adding one:

$$x + 0 = x$$
$$x + (y + 1) = (x + y) + 1.$$

This pair of equations shows that addition is obtained by primitive recursion from the functions $f(x) = x$ and $g(w, x, y) = w + 1$. These functions f and g are primitive recursive; f is the projection function I_1^1, and g is obtained by composition from successor and I_1^3. Putting these observations together, we can form a tree showing how addition is built up from the initial functions by composition and primitive recursion:

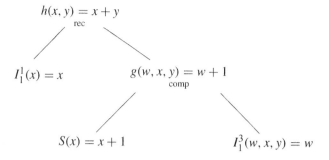

More generally, for any primitive recursive function h, we can use a labeled tree ("construction tree") to illustrate exactly how h is built up, as in the example of addition. At the top (root) vertex, we put h. At each minimal vertex (a leaf), we have an initial function: the successor function, a zero function, or a projection function. At each other vertex, we display either an application of composition or an application of primitive recursion.

An application of composition

$$h(\vec{x}) = f(g_1(\vec{x}), \ldots, g_n(\vec{x}))$$

can be illustrated in the tree by a vertex with $(n + 1)$-ary branching:

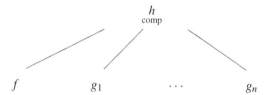

Here f must be an n-place function, and $g_1, \ldots, g_n$ must all have the same number of places as h.

An application of primitive recursion to obtain a $(k + 1)$-place function h

$$\begin{cases} h(\vec{x}, 0) = f(\vec{x}) \\ h(\vec{x}, y + 1) = g(h(\vec{x}, y), \vec{x}, y) \end{cases}$$

can be illustrated by a vertex with binary branching:

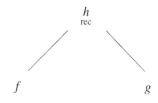

Note that g must have two more places than f, and one more place than h (e.g., if h is a two-place function, then g must be a three-place function and f must be a one-place function).

The $k = 0$ case, where a one-place function h is obtained by primitive recursion from a two-place function g by using the number m

$$\begin{cases} h(0) = m \\ h(x + 1) = g(h(x), x), \end{cases}$$

can be illustrated by a vertex with unary branching:

$$\begin{array}{c} h \\ \text{rec}(m) \end{array}$$

$$g$$

In both forms of primitive recursion ($k > 0$ and $k = 0$), the key feature is that the value of the function at a number $t + 1$ is somehow obtainable from its value at t. The role of g is to explain how.

Every primitive recursive function is total. We can see this by "structural induction." For the basis, all of the initial functions (the zero functions, the successor function, and the projections functions) are total. For the two inductive steps, we observe that composition of total functions yields a total function, and primitive recursion applied to total functions yields a total function. So for any primitive recursive function, we can work our way up its construction tree. At the leaves of the tree, we have total functions. And each time we move to a higher vertex, we still have a total function. Eventually, we come to the root at the top, and conclude that the function being constructed is total.

Next we want to build up a catalog of basic primitive recursive functions. These items in the catalog can then be used as "off the shelf" parts for later building up of other primitive recursive functions.

1. Addition $\langle x, y \rangle \mapsto x + y$ has already been shown to be primitive recursive.

The symbol "$\mapsto$" is read "maps to." The symbol gives us a very convenient way to name functions. For example, the squaring function can be named by the lengthy phrase "the function that given a number, squares it," which uses the pronoun "it" for the number. It is mathematically convenient to use a letter (such as x or t) in place of this pronoun. This leads us to the names "the function whose value at x is x^2" or "the function whose value at t is t^2." More compactly, these names can be written in symbols as "$x \mapsto x^2$" or "$t \mapsto t^2$." The letter x or t is a dummy variable; we can use any letter here.

2. Any constant function $\vec{x} \mapsto k$ can be obtained by applying composition k times to the successor function and the zero function $\vec{x} \mapsto 0$. For example, the three-place function that constantly takes the value 2 can be constructed by the following tree:

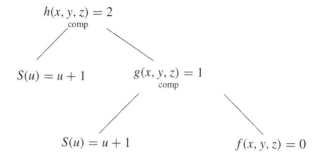

$$h(x, y, z) = 2$$

comp

$$S(u) = u + 1 \qquad g(x, y, z) = 1$$

comp

$$S(u) = u + 1 \qquad\qquad f(x, y, z) = 0$$

3. For multiplication $\langle x, y \rangle \mapsto x \times y$, we first observe that

$$x \times 0 = 0$$
$$x \times (y + 1) = (x \times y) + x.$$

This shows that multiplication is obtained by primitive recursion from the functions $x \mapsto 0$ and $\langle w, x, y \rangle \mapsto w + x$. The latter function is obtained by composition applied to addition and projection functions.

We can now conclude that any polynomial function with positive coefficients is primitive recursive. For example, we can see that the function $p(x, y) = x^2 y + 5xy + 3y^3$ is primitive recursive by repeatedly applying **1**, **2**, and **3**.

4. Exponentiation $\langle x, y \rangle \mapsto x^y$ is similar:

$$x^0 = 1$$
$$x^{y+1} = x^y \times x.$$

5. Exponentiation $\langle x, y \rangle \mapsto y^x$ is obtained from the preceding function by composition with projection functions. (The functions in items **4** and **5** are different functions; they assign different values to $\langle 2, 3 \rangle$. The fact that they coincide at $\langle 2, 4 \rangle$ is an accident.)

We should generalize this observation. For example, if f is primitive recursive, and g is defined by the equation

$$g(x, y, z) = f(y, 3, x, x)$$

then g is also primitive recursive, being obtained by composition from f and projection and constant functions. We will say in this situation that g is obtained from f by *explicit transformation*. Explicit transformation permits scrambling variables, repeating variables, omitting variables, and substituting constants.

6. The factorial function $x!$ satisfies the pair of recursion equations

$$0! = 1$$
$$(x + 1)! = x! \times (x + 1).$$

From this pair of equations, it follows that the factorial function is obtained by primitive recursion (by using 1) from the function $g(w, x) = w \cdot (x + 1)$. (See the example at the beginning of this chapter.)

7. The predecessor function $\text{pred}(x) = x - 1$ (except that $\text{pred}(0) = 0$) is obtained by primitive recursion from I_2^2:

$$\text{pred}\,(0) = 0$$
$$\text{pred}\,(x+1) = x.$$

This pair of equations leads to the tree:

pred
rec(0)

|

$I_2^2(w, x) = x$

8. Define the *proper subtraction* function $x \dotdiv y$ by the equation $x \dotdiv y = \max(x - y, 0)$. This function is primitive recursive:

$$x \dotdiv 0 = x$$
$$x \dotdiv (y+1) = \text{pred}(x \dotdiv y)$$

This pair of recursion equations yields the following construction tree:

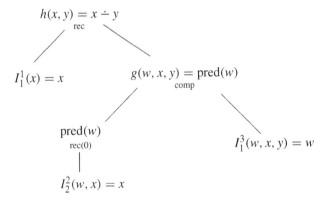

By the way, the symbol $\dotdiv$ is sometimes read as "monus."

9. Assume that f is primitive recursive, and define the functions s and p by the equations

$$s(\vec{x}, y) = \sum_{t<y} f(\vec{x}, t) \quad \text{and} \quad p(\vec{x}, y) = \prod_{t<y} f(\vec{x}, t)$$

(subject to the standard conventions for the empty sum $\sum_{t<0} f(\vec{x}, t) = 0$ and the empty product $\prod_{t<0} f(\vec{x}, t) = 1$). Then both s and p are primitive recursive. For p, we have the pair of equations:

$$p(\vec{x}, 0) = 1$$
$$p(\vec{x}, y+1) = p(\vec{x}, y) \cdot f(\vec{x}, y)$$

10. Define the function z by the equation

$$z(x) = \begin{cases} 1 & \text{if } x = 0 \\ 0 & \text{if } x > 0. \end{cases}$$

That is, the function z looks to see if its input is zero, and returns Yes (i.e., 1) if it is zero; otherwise, it returns No (i.e., 0). The function z is primitive recursive. We can see this from the equation $z(x) = 0^x$. More directly, we can see it from the equation $z(x) = 1 \dotminus x$. And even more directly, we can see it from the recursion equations

$$z(0) = 1$$
$$z(x+1) = 0$$

showing that z is obtained by primitive recursion (by using 1) from the function $g(w, x) = 0$.

$$
\begin{array}{c}
z \\
\text{rec(1)} \\
| \\
g(w, x) = 0
\end{array}
$$

11. In a similar vein, the function h that checks its two inputs x and y to see whether or not $x \le y$

$$h(x, y) = \begin{cases} 1 & \text{if } x \le y \\ 0 & \text{if otherwise} \end{cases}$$

is primitive recursive because $h(x, y) = z(x \dotminus y)$.

Items **10** and **11** can be reformulated in terms of relations (instead of functions). Suppose that R is a k-ary relation on the natural numbers, that is, R is some set of k-tuples of natural numbers: $R \subseteq \mathbb{N}^k$. We define R to be a *primitive recursive relation* if its characteristic function

$$C_R(x_1, \ldots, x_k) = \begin{cases} 1 & \text{if } \langle x_1, \ldots, x_k \rangle \in R \\ 0 & \text{if otherwise} \end{cases}$$

is a primitive recursive function. For example, item **11** states that the ordering relation $\{\langle x, y \rangle \mid x \le y\}$ is a primitive recursive binary relation. And item **10** states that $\{0\}$ is a primitive recursive unary relation.

From composition, we derive the *substitution rule*: If Q is an n-ary primitive recursive relation, and $g_1, \ldots, g_n$ are k-place primitive recursive functions, then the k-ary relation

$$\{\vec{x} \mid \langle g_1(\vec{x}), \ldots, g_n(\vec{x}) \rangle \in Q\}$$

is primitive recursive because its characteristic function is obtained from C_Q and $g_1, \ldots, g_n$ by composition.

From a relation Q, we can form its *complement* $\overline{Q}$:

$$\overline{Q} = \{\vec{x} \mid \vec{x} \text{ is not in } Q\}$$

From two k-ary relations Q and R (for the same k), we can form their *intersection*,

$$Q \cap R = \{\vec{x} \mid \text{both } \vec{x} \in Q \text{ and } \vec{x} \in R\}$$

and their *union*

$$Q \cup R = \{\vec{x} \mid \text{either } \vec{x} \in Q \text{ or } \vec{x} \in R \text{ or both}\}.$$

We can streamline the notation slightly by writing, instead of $\vec{x} \in Q$, simply $Q(\vec{x})$. In this notation,

$$\overline{Q} = \{\vec{x} \mid \text{not } Q(\vec{x})\},$$
$$Q \cap R = \{\vec{x} \mid \text{both } Q(\vec{x}) \text{ and } R(\vec{x})\},$$
$$Q \cup R = \{\vec{x} \mid \text{either } Q(\vec{x}) \text{ or } R(\vec{x}) \text{ or both}\}.$$

The following theorem assures us that these constructions preserve primitive recursiveness. That is, when applied to primitive recursive relations, they produce primitive recursive relations. This theorem will be useful in extending our supply of primitive recursive relations and functions.

Theorem: *Assume that Q and R are k-ary primitive recursive relations. Then the following relations are also primitive recursive:*

(a) *The complement $\overline{Q}$ of Q:*

$$\overline{Q} = \{\vec{x} \mid \text{not } Q(\vec{x})\}$$

(b) *The intersection $Q \cap R$ of Q and R:*

$$Q \cap R = \{\vec{x} \mid \text{both } Q(\vec{x}) \text{ and } R(\vec{x})\}$$

(c) *The union $Q \cup R$ of Q and R:*

$$Q \cup R = \{\vec{x} \mid \text{either } Q(\vec{x}) \text{ or } R(\vec{x}) \text{ or both}\}$$

Proof.

(a)

$$C_{\overline{Q}}(\vec{x}) = z(C_Q(\vec{x}))$$

where z is the function from **10**. That is, $C_{\overline{Q}}$ is obtained by composition from functions known to be primitive recursive. The other parts are proved similarly; we need to make the characteristic function from primitive recursive parts.

(b)

$$C_{Q \cap R}(\vec{x}) = C_Q(\vec{x}) \cdot C_R(\vec{x})$$

(c)

$$C_{Q \cup R}(\vec{x}) = \text{pos}[C_Q(\vec{x}) + C_R(\vec{x})]$$

where pos is the function from Exercise 5. ⊣

For example, we can apply this theorem to conclude that $>$ and $=$ are primitive recursive relations:

12. The relation $\{\langle x, y \rangle \mid x > y\}$ is primitive recursive because it is the complement of the $\leq$ relation from item **11**.

13. The relation $\{\langle x, y \rangle \mid x = y\}$ is primitive recursive because it is the intersection of the $\leq$ and the $\geq$ relations, and $\geq$ is obtained from $\leq$ by explicit transformation.

It follows from item **13** and the substitution rule that for any primitive recursive function f, its graph

$$\{\langle \vec{x}, y \rangle \mid f(\vec{x}) = y\}$$

is a primitive recursive relation.

Definition by cases: *Assume that Q is a primitive recursive k-ary relation, and that f and g are primitive recursive k-place functions. Then the function h defined by the equation*

$$h(\vec{x}) = \begin{cases} f(\vec{x}) & \text{if } Q(\vec{x}) \\ g(\vec{x}) & \text{if not } Q(\vec{x}) \end{cases}$$

is also primitive recursive.

Proof. $h(\vec{x}) = f(\vec{x}) \cdot C_Q(\vec{x}) + g(\vec{x}) \cdot C_{\overline{Q}}(\vec{x}).$ ⊣

This result can be extended to more than two cases; see Exercise 12. For example, we might want to handle an equation of the form

$$h(\vec{x}) = \begin{cases} f_1(\vec{x}) & \text{if } Q(\vec{x}) \text{ and } R(\vec{x}) \\ f_2(\vec{x}) & \text{if } Q(\vec{x}) \text{ and not } R(\vec{x}) \\ f_3(\vec{x}) & \text{if } R(\vec{x}) \text{ and not } Q(\vec{x}) \\ f_4(\vec{x}) & \text{if neither } Q(\vec{x}) \text{ nor } R(\vec{x}) \end{cases}$$

or one of the form

$$h(\vec{x}) = \begin{cases} f_1(\vec{x}) & \text{if } Q_1(\vec{x}) \\ f_2(\vec{x}) & \text{if } Q_2(\vec{x}) \\ \cdots & \cdots \\ f_9(\vec{x}) & \text{if } Q_9(\vec{x}) \\ f_{10}(\vec{x}) & \text{if none of the above} \end{cases}$$

in a situation in which it is known that no two of $Q_1, \ldots, Q_9$ can hold simultaneously. Moreover, from a k-ary relation Q, we can form

$$\{\langle x_1, \ldots, x_{k-1}, y \rangle \mid \text{ for every } t < y, \langle x_1, \ldots, x_{k-1}, t \rangle \in Q\},$$

which can be written in better notation as

$$\{\langle \vec{x}, y \rangle \mid (\forall t < y) \, Q(\vec{x}, t)\},$$

where the symbol $\forall$ is read "for all." In the same spirit, we can form

$$\{\langle x_1, \ldots, x_{k-1}, y \rangle \mid \text{ for some } t < y, \langle x_1, \ldots, x_{k-1}, t \rangle \in Q\},$$

which is better written as

$$\{\langle \vec{x}, y \rangle \mid (\exists t < y) \, Q(\vec{x}, t)\},$$

where the symbol $\exists$ is read as "there exists . . . such that."

Again, these constructions preserve primitive recursiveness:

Theorem: *Assume that Q is a $(k + 1)$-ary primitive recursive relation. Then the following relations are also primitive recursive:*

(a)

$$\{\langle \vec{x}, y \rangle \mid (\forall t < y) \, Q(\vec{x}, t)\}$$

(b)

$$\{\langle \vec{x}, y \rangle \mid (\exists t < y) \, Q(\vec{x}, t)\}$$

Proof.

(a) The value of the characteristic function at $\langle \vec{x}, y \rangle$ is

$$\prod_{t < y} C_Q(\vec{x}, t).$$

Apply item **9**.

(b) The value of the characteristic function at $\langle \vec{x}, y \rangle$ is

$$\text{pos} \left[\sum_{t < y} C_Q(\vec{x}, t) \right]$$

where pos is the function from Exercise 5. This is primitive recursive by item **9** and Exercise 5.

⊣

For example, we can apply these results to show that the relation

$$\{ \langle x, y \rangle \mid (\exists q < y + 1)[x \cdot q = y] \}$$

is primitive recursive. We do this by looking at the way the above line is written, and then filling in the details. First of all, the ternary relation

$$R_1(x, y, q) \iff x \cdot q = y$$

is obtained from the equality relation by substituting the functions $\langle x, y, q \rangle \mapsto x \cdot q$ and I_2^3. Secondly, an application of the preceding theorem then shows that the ternary relation

$$R_2(x, y, z) \iff (\exists q < z)[x \cdot q = y]$$

is primitive recursive. Finally, we apply substitution:

$$(\exists q < y + 1)[x \cdot q = y] \iff R_2(x, y, y + 1).$$

In short, we can show that this relation is primitive recursive by examining the syntactical form of its definition and verifying that it has been built up by using only pieces that are known to be primitive recursive.

14. The divisibility relation $x \mid y$, that is, the relation

$$\{ \langle x, y \rangle \mid x \text{ divides } y \text{ with } 0 \text{ remainder} \},$$

is primitive recursive. (Here we adopt the convention that 0 divides itself, but it does not divide any positive integer.) This is because

$$\begin{aligned}
x \mid y &\iff \text{for some quotient } q, \text{ we have } x \cdot q = y \\
&\iff (\exists q \leq y)[x \cdot q = y] \\
&\iff (\exists q < y + 1)[x \cdot q = y].
\end{aligned}$$

That is, the relation we examined in the foregoing example is nothing but the divisibility relation!

In effect, we are building up a certain *language* such that any function or relation definable in the language is guaranteed to be primitive recursive. (For divisibility, the crucial fact was that the expression "$(\exists q < y + 1)[x \cdot q = y]$" belonged to this language.) This language includes the following:

- Variables: The projection functions are primitive recursive.
- Constants (numerals): The constant functions are primitive recursive.
- Function symbols: We can use symbols for any primitive recursive function in the list we are building up $(+, \times, \div, \ldots$, with more to come).
- Combinations: $\sum_{x<y}, \prod_{x<y}$, with more to come.
- Relation symbols: We can use symbols for any primitive recursive relation in the list we are building up $(\leq, =, |, \ldots$, with more to come).
- More combinations: "not," "and," "or" can be applied to relations.
- Bounded "quantifiers": $\forall x < y$ and $\exists x < y$. (The upper bound y is needed here.)

We have theorems assuring us that functions or relations expressible in this language are certain to be primitive recursive.

For example, we next add the set of primes (as a unary relation) to our list:

15. The set $\{2, 3, 5, \ldots\}$ of prime natural numbers (as a unary relation on $\mathbb{N}$) is primitive recursive. To see this, observe that

$$x \text{ is prime} \iff 1 < x \text{ and } (\forall u < x)(\forall v < x)[uv \neq x],$$

and the right-hand side is written within the language available to us.

2.1.1 Bounded Search

The search operator (often called minimalization or the μ-operator) provides a useful way of defining a function in terms of a "search" for the first time a given condition is satisfied.

Definition: For a $(k + 1)$-ary relation P, the number $(\mu t < y)P(\vec{x}, t)$ is defined by the equation:

$$(\mu t < y)P(\vec{x}, t) = \begin{cases} \text{the least } t \text{ such that } t < y \text{ and } P(\vec{x}, t), \text{ if any} \\ y \quad \text{if there is no such } t \end{cases}$$

For example, if we let

$$f(x, y) = \mu t < y[t \text{ is prime and } x < t]$$

then $f(6, 4) = 4$ and $f(6, 8) = f(6, 800) = 7$.

Theorem: *If P is a primitive recursive relation, then the function*

$$f(\vec{x}, y) = (\mu t < y)P(\vec{x}, t)$$

is a primitive recursive function.

Proof. We will apply primitive recursion. Trivially $f(\vec{x}, 0) = 0$, so there is no problem here. The problem to see how $f(\vec{x}, y+1)$ (call this b) depends on $f(\vec{x}, y)$ (call this a):

- If $a < y$, then $b = a$. (The search below y succeeded.)
- If $a = y$ and $P(\vec{x}, y)$, then $b = y$.
- Otherwise, $b = y + 1$.

Thus, if we define

$$g(a, \vec{x}, y) = \begin{cases} a & \text{if } a < y \\ y & \text{if } a \not< y \text{ and } P(\vec{x}, y) \\ y+1 & \text{if } a \not< y \text{ and not } P(\vec{x}, y), \end{cases}$$

then f is obtained by primitive recursion from the functions $\vec{x} \mapsto 0$ and g. Because P is a primitive recursive relation, it follows that g is primitive recursive (by definition-by-cases), and hence f is primitive recursive.   ⊣

There is another proof of this theorem, which relies on the following remarkable equation:

$$(\mu t < y) P(\vec{x}, t) = \sum_{u < y} \prod_{t \leq u} C_{\overline{P}}(\vec{x}, t)$$

A related search operator is bounded *maximalization*. Define the $\overline{\mu}$-operator as follows:

$$(\overline{\mu} t \leq y) P(\vec{x}, t) = \begin{cases} \text{the largest number } t \text{ such that } t \leq y \text{ and } P(\vec{x}, t), \text{ if any} \\ 0 \quad \text{if there is no such } t \end{cases}$$

Theorem: *If P is a primitive recursive relation, then the function*

$$f(\vec{x}, y) = (\overline{\mu} t \leq y) P(\vec{x}, t)$$

is a primitive recursive function.

Proof.

$$(\overline{\mu} t \leq y) P(\vec{x}, t) = y \div (\mu s < y) P(\vec{x}, y \div s).$$

This equation captures the idea of searching down from y.   ⊣

Euclid observed that the set of prime numbers is unbounded. Hence the function

$$h(x) = \text{the smallest prime number larger than } x$$

is total. It is also primitive recursive because

$$h(x) = \mu t < (x! + 2)[t \text{ is prime and } x < t].$$

The upper bound $x! + 2$ suffices because for any prime factor p of $x! + 1$, we have $x < p \leq x! + 1$. So any search for a prime larger than x need go no further than $x! + 1$.

Digression: There is an interesting result in number theory here. "Bertrand's postulate" states that for any $x > 3$, there will always be a prime number p with $x < p < 2x - 2$. (Bertrand's postulate implies that in the previous paragraph, it suffices to use simply $h(x) = \mu t < (2x+3)[t$ is prime and $x < t]$.) In 1845, the French mathematician Joseph Bertrand, using prime number tables, verified this statement for x below three million. Then in 1850, the Russian P. L. Chebyshev (Tchebychef) proved the result in general. In 1932, the Hungarian Paul Erdős gave a better proof, which can now be found in undergraduate number theory textbooks. The origin of

> Chebyshev said it
> So I'll say it again
> There's always a prime
> Between N and $2N$

is unknown, which may be just as well.

Define p_x to be the $(x + 1)$st prime number, so that

$$p_0 = 2, \quad p_1 = 3, \quad p_2 = 5, \quad p_3 = 7, \quad p_4 = 11,$$

and so forth. In other words, p_x is the xth odd prime, except that $p_0 = 2$. (The prime number theorem tells us that p_x grows at a rate something like $x \ln x$, but that is beside the point.)

16. The function $x \mapsto p_x$ is primitive recursive because we have the recursion equations

$$p_0 = 2$$
$$p_{x+1} = h(p_x),$$

where h is the above function that finds the next prime.

It is easy to see that we always have $x + 1 < p_x$; a formal proof can use induction.

We will need methods for encoding a string of numbers by a single number. One method that is conceptually simple uses powers of primes. We define the "bracket notation" as follows.

$$[\,] = 1$$
$$[x] = 2^{x+1}$$
$$[x, y] = 2^{x+1}3^{y+1}$$
$$[x, y, z] = 2^{x+1}3^{y+1}5^{z+1}$$
$$\cdots$$
$$[x_0, x_1, \ldots, x_k] = 2^{x_0+1}3^{x_1+1}\cdots p_k^{x_k+1}$$

For example, $[2, 1] = 72$ and $[2, 1, 0] = 360$. Clearly, for any one value of k, the function

$$\langle x_0, x_1, \ldots, x_k \rangle \mapsto [x_0, x_1, \ldots, x_k]$$

is a primitive recursive $(k + 1)$-place function. But encoding is useless, unless we can *decode*. (The "fundamental theorem of arithmetic" is the statement that every positive integer has a factorization into primes, unique up to order. For decoding, we are implicitly exploiting the uniqueness of prime factorization.) Item **17** will give a primitive recursive decoding function.

Digression: Using powers of primes is by no means the only way to encode a string of numbers. It is a very convenient method for our present purposes, but there are a number of other methods. Here is a very different approach:

$$\langle x_0, x_1, \ldots, x_k \rangle \mapsto 1\underbrace{00\cdots0}_{x_0}1\underbrace{00\cdots0}_{x_1}1\cdots\cdots1\underbrace{00\cdots0}_{x_k}{}_{\text{two}}$$

That is, a sequence of length n can be coded by the number whose binary representation has n 1's. The number of 0's that follow the ith 1 in the representation corresponds to the ith component in the sequence.

Here are some examples:

$$\langle 0, 3, 2 \rangle \mapsto 11000100_{\text{two}} = 188$$
$$\langle 2, 1, 0 \rangle \mapsto 100101_{\text{two}} = 37$$
$$\langle\ \rangle \mapsto 0_{\text{two}} = 0$$
$$\langle 7 \rangle \mapsto 10000000_{\text{two}} = 128$$
$$\langle 0, 0, 0, 0 \rangle \mapsto 1111_{\text{two}} = 31$$

These values can be compared with the values yielded by the bracket notation: $[0, 3, 2] = 2^1 \cdot 3^4 \cdot 5^3 = 20, 250, [2, 1, 0] = 2^3 \cdot 3^2 \cdot 5^1 = 360, [\ ] = 1, [7] = 2^8 = 256, [0, 0, 0, 0] = 2 \cdot 3 \cdot 5 \cdot 7 = 210.$

In particular, suppose we want to encode a sequence of *two* numbers. This method yields

$$\langle m, n \rangle \mapsto 1\underbrace{00\cdots0}_{m}1\underbrace{00\cdots0}_{n}{}_{\text{two}} \qquad = 2^{m+n+1} + 2^n = 2^n(2^{m+1} + 1).$$

The bracket notation yields simply $[m, n] = 2^{m+1} \cdot 3^{n+1}$. Both of these "pairing functions" have the feature that they grow exponentially as m and n increase.

Interestingly, there are *polynomial* pairing functions, and here is one:

$$J(m, n) = \frac{1}{2}((m + n)^2 + 3m + n).$$

The function J is one-to-one, so the pair $\langle m, n \rangle$ is recoverable from the value $J(m, n)$. In fact the function J maps $\mathbb{N} \times \mathbb{N}$ one-to-one *onto* $\mathbb{N}$.

And where does J come from? Here is a clue. Calculate $J(m, n)$ for all small values of m and n, say $m + n \le 4$. Then make a chart in the plane, by placing the number $J(m, n)$ at the point in the plane with coordinates $\langle m, n \rangle$. Check if a pattern is emerging.

17. There is a primitive recursive two-place "decoding" function, whose value at $\langle x, y \rangle$ is written $(x)_y$, with the property that whenever $y \leq k$,

$$([x_0, x_1, \ldots, x_k])_y = x_y.$$

That is,

(code for a sequence)$_y$ = the $(y + 1)$st term of the sequence.

For example, $(72)_0 = 2$ and $(72)_1 = 1$ because $72 = [2, 1]$.

First, observe that the exponent of a prime q in the factorization of a positive integer x is

$$\mu e \left(q^{e+1} \nmid x \right),$$

the smallest e for which $e + 1$ would be too much. We can bound the search at x because if $q^e \mid x$, then $e < q^e \leq x$. That is, the exponent of q in the factorization of x is

$$(\mu e < x) \left(q^{e+1} \nmid x \right).$$

Now suppose that prime q is p_y. We define

$$(x)_y^* = (\mu e < x) \left(p_y^{e+1} \nmid x \right)$$

so that $(x)_y^*$ is the exponent of p_y in the prime factorization of x.

Secondly, for our decoding function, we need one *less* than the exponent of the prime p_y in the factorization of the sequence code. Accordingly, we define

$$(x)_y = (x)_y^* \dotminus 1 = (\mu e < x) \left(p_y^{e+1} \nmid x \right) \dotminus 1.$$

The right-hand side of this equation is written in our language, so the function is primitive recursive. The function tests powers of p_y until it finds the largest one in the factorization of x, and then it backs down by 1. If p_y does not divide x, then $(x)_y = 0$, harmlessly enough. Also $(0)_y = 0$, but for a different reason.

18. Say that y is a *sequence number* if either $y = [\,]$ or $y = [x_0, x_1, \ldots, x_k]$ for some k and some $x_0, x_1, \ldots, x_k$. For example, 1 is a sequence number but 50 is not. The set of sequence numbers is primitive recursive; see Exercise 14. The set of sequence numbers starts off as $\{1, 2, 4, 6, 8, 12, \ldots\}$.

19. There is a primitive recursive function lh such that

$$\text{lh}[x_0, x_1, \ldots, x_k] = k + 1.$$

For example, $\text{lh}(360) = 3$. Here "lh" stands for "length." We define

$$\text{lh}(x) = (\mu k < x) \left(p_k \nmid x \right).$$

Thus, for example, $\text{lh}(50) = 1$.

It is apparent from its definition that this function is primitive recursive. The upper bound on the μ search is adequate because if $p_{k-1} \mid x$, then $(k-1) + 1 < p_{k-1} \le x$. If s is a sequence number of positive length, then

$$(s)_{\text{lh}(s) \dot- 1}$$

will be the *last* component of the sequence.

20. There is a two-place primitive recursive function whose value at $\langle x, y \rangle$ is called the *restriction* of x to y, written $x \restriction y$, with the property that whenever $y \le k + 1$ then

$$[x_0, x_1, \ldots, x_k] \restriction y = [x_0, x_1, \ldots, x_{y-1}].$$

That is, the restriction of x to y gives us the first y components of the sequence. We define

$$x \restriction y = \prod_{i<y} p_i^{(x)_i^*}.$$

For example, if s is a sequence number, then $s \restriction (\text{lh}(s) \dot- 1)$ will encode the result of deleting the *last* item in the sequence, if any.

21. There is a two-place primitive recursive function whose value at $\langle x, y \rangle$ is called the *concatenation* of x and y, written $x * y$, with the property that whenever x and y are sequence numbers, then $x * y$ is the sequence number of length $\text{lh}(x) + \text{lh}(y)$ whose components are first the components of x and then the components of y. We define

$$x * y = x \cdot \prod_{i<\text{lh}(y)} p_{i+\text{lh}(x)}^{(y)_i^*}.$$

For example, $72 * 72 = [2, 1, 2, 1] = 441,000$. If s is a sequence number, then $s * [x]$ will encode the result of adjoining x to the *end* of the sequence.

22. We can also define a "capital asterisk" operation. Let

$$\textstyle\bigast_{t<y} a_t = a_0 * a_1 * \cdots * a_{y-1}$$

(grouped to the left). If f is a primitive recursive $(k + 1)$-place function, then so is the function whose value at $\langle \vec{x}, y \rangle$ is $\bigast_{t<y} f(\vec{x}, t)$, as can be seen from the pair of recursion equations:

$$\textstyle\bigast_{t<0} f(\vec{x}, t) = 1$$

$$\textstyle\bigast_{t<y+1} f(\vec{x}, t) = \bigast_{t<y} f(\vec{x}, t) * f(\vec{x}, y)$$

For any $(k + 1)$-place function f, we define $\bar{f}$ by the equation

$$\bar{f}(\vec{x}, y) = [f(\vec{x}, 0), f(\vec{x}, 1), \ldots, f(\vec{x}, y - 1)]$$

so that the number $\bar{f}(\vec{x}, y)$ encodes y values of f, namely the values $f(\vec{x}, t)$ for all $t < y$. For example, $\bar{f}(\vec{x}, 0) = [\ ] = 1$, encoding 0 values. And $\bar{f}(\vec{x}, 2) = [f(\vec{x}, 0), f(\vec{x}, 1)]$. Clearly $\bar{f}(\vec{x}, y)$ is always a sequence number of length y.

23. If f is primitive recursive, then so is $\bar{f}$ because

$$\bar{f}(\vec{x}, y) = \prod_{i < y} p_i^{f(\vec{x}, i) + 1}.$$

Now suppose we have a $(k + 2)$-place function g. Then there exists a unique function $(k + 1)$-place f satisfying the equation

$$f(\vec{x}, y) = g(\bar{f}(\vec{x}, y), \vec{x}, y)$$

for all $\vec{x}$ and y. For example,

$$f(\vec{x}, 0) = g([\,], \vec{x}, 0) = g(1, \vec{x}, 0)$$
$$f(\vec{x}, 1) = g([f(\vec{x}, 0)], \vec{x}, 1)$$
$$f(\vec{x}, 2) = g([f(\vec{x}, 0), f(\vec{x}, 1)], \vec{x}, 2)$$

and so forth. The function f is determined recursively; we can find $f(\vec{x}, y)$ after we know $f(\vec{x}, t)$ for all $t < y$.

24. Assume that g is a primitive recursive $(k + 2)$-place function, and let f be the unique $(k + 1)$-place function for which

$$f(\vec{x}, y) = g(\bar{f}(\vec{x}, y), \vec{x}, y)$$

for all $\vec{x}$ and y. Then f is also primitive recursive.

To see that f is primitive recursive, we first examine $\bar{f}$. We have the pair of recursion equations

$$\bar{f}(\vec{x}, 0) = 1$$
$$\bar{f}(\vec{x}, y + 1) = \bar{f}(\vec{x}, y) * [g(\bar{f}(\vec{x}, y), \vec{x}, y)]$$

from which we see that $\bar{f}$ is primitive recursive. Secondly, the primitive recursiveness of f itself follows from the equation

$$f(\vec{x}, y) = g(\bar{f}(\vec{x}, y), \vec{x}, y)$$

once we know that $\bar{f}$ is primitive recursive.

The definition of primitive recursion involved defining the value of a function in terms of its immediately preceding value. Item **24** shows that we get an added bonus: the value of a function can be defined in terms of *all* its preceding values.

At this point, we have seen that many of the everyday functions on the natural numbers are primitive recursive. But the class of primitive recursive functions does not include *all* of the functions on $\mathbb{N}$ that one would regard as effectively calculable. W. Ackermann showed how to construct an effectively calculable function that grows faster than any primitive recursive function. Also, we can "diagonalize out" of the primitive recursive functions. In rough outline, here is how that would go: Any primitive recursive function is determined by tree, showing how it is built up from initial functions by the use of composition and primitive recursion. We can, with some effort, code such trees by natural numbers. The "universal" function

$$\Psi(x, y) = \begin{cases} f(x) & \text{if } y \text{ codes a tree for a one-place primitive recursive function } f \\ 0 & \text{otherwise} \end{cases}$$

is effectively calculable (and total). But $\Psi(x, x)+1$ and $1 \mathbin{\dot-} \Psi(x, x)$ are total effectively calculable functions that cannot be primitive recursive. (See also page 19.)

2.2 Search Operation

We obtain the class of general recursive partial functions by allowing functions to be built up by use of search (in addition to composition and primitive recursion). Search (also called minimalization) corresponds to an unbounded μ-operator. For a $(k + 1)$-place partial function g, we define

$$\mu y[g(\vec{x}, y) = 0] = \begin{cases} \text{the least number } y \text{ such that both } g(\vec{x}, y) = 0 \text{ and} \\ \quad \text{for all } t \text{ less than } y, \text{ the value } g(\vec{x}, t) \text{ is defined} \\ \quad \text{and is nonzero, if there is any such } y \\ \text{undefined, if there is no such } y. \end{cases}$$

This quantity may be undefined for some (or all) values of $\vec{x}$, even if g happens to be a total function.

Example: Assume that we know the following pieces of information about the function g:

$$g(0, 0) = 7 \quad g(0, 1) = 0$$
$$g(1, 0) \uparrow \quad g(1, 1) = 0$$

Then $\mu y[g(0, y) = 0]$ is 1, and $\mu y[g(1, y) = 0]$ is undefined.

A k-place partial function h is said to be obtained from g by search if the equation

$$h(\vec{x}) = \mu y[g(\vec{x}, y) = 0]$$

holds for all $\vec{x}$, with the usual understanding that for an equation to hold, either both sides are undefined, or both sides are defined and are equal.

Then we say that a partial function is *general recursive* if it can be built up from the zero, successor, and projection functions, where we are allowed to use composition, primitive recursion, and search.

The collection of general recursive partial functions includes all of the primitive recursive functions (which are all total), and more. As an extreme example, the one-place empty function (i.e., the function with empty domain) is a general recursive partial function; it is obtained by search from the constant function $g(x, y) = 3$.

9A. If f is a general recursive partial function, then so are the functions s and p:

$$s(\vec{x}, y) = \sum_{t<y} f(\vec{x}, t) \quad \text{and} \quad p(\vec{x}, y) = \prod_{t<y} f(\vec{x}, t)$$

For any particular $\vec{x}$, these functions are defined either for all y, or for a finite initial segment of the natural numbers.

We define a relation R to be a *general recursive relation* if its characteristic function C_R (which by definition is always total) is a general recursive function. As a special case of search, whenever R is a $(k+1)$-ary general recursive relation, then the k-place function h defined by the equation

$$h(\vec{x}) = \mu y\, R(\vec{x}, y)$$

is a general recursive partial function.

We again have a *substitution rule*: Whenever Q is an n-ary general recursive relation, and $g_1, \ldots, g_n$ are k-place *total* general recursive functions, then the k-ary relation

$$\{\vec{x} \mid \langle g_1(\vec{x}), \ldots, g_n(\vec{x})\rangle \in Q\}$$

is general recursive because its characteristic function is obtained from C_Q and $g_1, \ldots, g_n$ by composition. But this does not necessarily hold if the g_i functions are nontotal. In that case, composition does not give us the full C_Q, but a nontotal subfunction of it.

For example, for any *total* general recursive function f, its graph

$$\{\langle \vec{x}, y\rangle \mid f(\vec{x}) = y\}$$

is a general recursive relation. (Similarly, the graph of any primitive recursive function will be a primitive recursive relation.) We will see later that this can fail in the case of a nontotal function.

Theorem:
(d) *If Q and R are k-ary general recursive relations, then so are $\overline{Q}$, $Q \cap R$, and $Q \cup R$.*
(e) *If Q is a $(k+1)$-ary general recursive relation, then so are the relations*

$$\{\langle \vec{x}, y\rangle \mid (\forall t < y)\, Q(\vec{x}, t)\} \quad \text{and} \quad \{\langle \vec{x}, y\rangle \mid (\exists t < y)\, Q(\vec{x}, t)\}.$$

The proof is unchanged.

Definition-by-cases continues to hold, but we need to be more careful with its proof. Suppose that g is a k-place general recursive partial function, and that Q is a k-ary general recursive relation. Define g^Q by the equation

$$g^Q(\vec{x}) = \begin{cases} g(\vec{x}) & \text{if } Q(\vec{x}) \\ 0 & \text{if not } Q(\vec{x}). \end{cases}$$

Then g^Q is also a general recursive partial function. But we cannot write simply $g^Q(\vec{x}) = g(\vec{x}) \cdot C_Q(\vec{x})$ because there may be some $\vec{x}$ that are not in the domain of g (so the right-hand side will be undefined) and not in Q (so the left-hand side will be 0). Instead, we can first use primitive recursion to construct the function

$$G(\vec{x}, 0) = 0$$
$$G(\vec{x}, y + 1) = g(\vec{x}),$$

which is like g except that it has a "on–off switch." Then we have the equation

$$g^Q(\vec{x}) = G(\vec{x}, C_Q(\vec{x})).$$

showing that g^Q is a general recursive partial function.

Now if we also have another k-place general recursive partial function f, and we define

$$h(\vec{x}) = \begin{cases} f(\vec{x}) & \text{if } Q(\vec{x}) \\ g(\vec{x}) & \text{if not } Q(\vec{x}) \end{cases}$$

then h is a general recursive partial function because $h(\vec{x}) = f^Q(\vec{x}) + g^{\overline{Q}}(\vec{x})$.

24A. Assume that g is a general recursive partial $(k + 2)$-place function, and let f be the unique $(k + 1)$-place function for which

$$f(\vec{x}, y) = g(\bar{f}(\vec{x}, y), \vec{x}, y)$$

for all $\vec{x}$ and y. (If g is nontotal, then it is possible that for some values of $\vec{x}$, the quantity $f(\vec{x}, y)$ will be defined only for finitely many y's.) Then f is also a general recursive partial function.

The proof is as before.

It was argued earlier that the collection of primitive recursive functions cannot contain all of the effectively calculable total functions. But Church's thesis implies that the collection of general recursive partial functions *does* contain all of them, as well as the effectively calculable nontotal functions. As indicated informally on page 20, it is not possible to "diagonalize out" of the collection of general recursive partial functions.

Exercises

0. Do you understand primitive recursion? Are you positive? If you are positive, go to Exercise 1.

1. Subtract 1. Go to Exercise 0.

2. Give an example of a nontotal function g such that the function h obtained from g by search

$$h(x) = \mu y[g(x, y) = 0]$$

is total.

3. Give a construction tree in full for multiplication (item **3**).

4. Show that the squaring function $f(x) = x^2$ is primitive recursive by giving a construction tree showing in detail how it can be built up from initial functions by the use of composition and primitive recursion. (At the leaves of the tree, you must have only initial functions; e.g., if you want to use addition, you must construct it.)

5. Show that the function

$$\mathrm{pos}(x) = \begin{cases} 1 & \text{if } x > 0 \\ 0 & \text{if } x = 0 \end{cases}$$

is primitive recursive by giving a construction tree.

6. Show that the parity function

$$C_{\mathrm{odd}}(x) = \begin{cases} 1 & \text{if } x \text{ is odd} \\ 0 & \text{if } x \text{ is even} \end{cases}$$

is primitive recursive by giving a construction tree.

7. Show that the function $\langle x, y \rangle \mapsto |x - y|$ is primitive recursive.

8. Show that the function $\langle x, y \rangle \mapsto \max(x, y)$ is primitive recursive.

9. Show that the function $\langle x, y \rangle \mapsto \min(x, y)$ is primitive recursive.

10. Show that there is a primitive recursive function div such that whenever $y > 0$, then

$$\mathrm{div}(x, y) = \lfloor x/y \rfloor.$$

(Here $\lfloor z \rfloor$ is the largest natural number that is $\leq z$, i.e., the result of rounding z down to a natural number.)

11. Show that there is a primitive recursive function rm such that whenever $y > 0$, then

$$\mathrm{rm}(x, y) = \text{the remainder when } x \text{ is divided by } y.$$

12. Extend definition by cases (pages 37 and 48) to definition by many (mutually exclusive) cases.

13. Use Bertrand's postulate to show (by induction) that $p_x \leq 2^{x+1}$, and that equality holds only for $x = 0$.

14. Prove item **18**: The set of sequence numbers is primitive recursive.

15. Show that $(x)_y = (\overline{\mu}e \leq x) \, p_y^{e+1} \mid x$.

16. Show that $\mathrm{lh}(x) = (\overline{\mu}t \leq x) \, p_{t \dot- 1} \mid x$, for a sequence number x.

17. Assume that R is a finite k-ary relation on $\mathbb{N}$ (i.e., R is a finite subset of $\mathbb{N}^k$). Show that R is primitive recursive.

18. Assume that f is an eventually constant one-place function (i.e., there is some m such that $f(x + 1) = f(x)$ for all $x \leq m$). Show that f is primitive recursive.

19. Show that the function $g(x) = \lceil \sqrt{x} \rceil$ is primitive recursive. (Here $\lceil z \rceil$ is the result of rounding z *up* to a natural number.)

20. (a) Assume that f is a primitive recursive one-place function that is strictly increasing (i.e., $f(x+1) > f(x)$ for all x). Show that the range of f is a primitive recursive set.

(b) Assume that g is a primitive recursive one-place function that is nondecreasing (i.e., $g(x+1) \geq g(x)$ for all x) and unbounded. Show that the range of g is a general recursive set.

21. Assume that h is a finite k-place function (i.e., the domain of h consists of only finitely many k-tuples). Show that h is a general recursive partial function.

22. Is 3 a sequence number? What is $\text{lh}(3)$? Find $(1 * 3) * 6$ and $1 * (3 * 6)$.

23. Show that $*$ is associative on sequence numbers. That is, show that if r, s, and t are sequence numbers, then $(r * s) * t = r * (s * t)$.

24. Establish the following facts.

(a) $x + 1 < p_x$.

(b) $(y)_k \leq y$, and equality holds iff $y = 0$.

(c) $\text{lh}\, x \leq x$, and equality holds iff $x = 0$.

(d) $x \upharpoonright i \leq x$ if $x > 0$.

(e) $\text{lh}(x \upharpoonright i)$ is the smaller of i and $\text{lh}\, x$.

3 Programs and Machines

In this chapter, we focus on another way of formalizing the concept of effective calculability, namely register-machine programs. Our first goal is to show that all general recursive partial functions are also computable by register machines. This fact will allow us to apply our work in Chapter 2 to see that a great many everyday functions are register-machine computable. Using this, we will be able to construct a universal program, that is, a program to compute the partial function $\Phi(w, x) = $ the result of applying the program coded by w to the input x.

3.1 Register Machines

Recall from Chapter 1 that a register-machine program is a finite sequence of instructions of the following types:

- "Increment r," I r (where r is a numeral for a natural number): The effect of this instruction is to increase the contents of register r by 1. The machine then proceeds to the next instruction in the program (if any).
- "Decrement r," D r (where r is a numeral for a natural number): The effect of this instruction depends on the contents of register r. If that number is nonzero, it is decreased by 1 and the machine proceeds *not* to the next instruction, but to the following one. But if the number in register r is zero, the machine simply proceeds to the next instruction. In summary, the machines tries to decrement register r, and if it is successful, then it skips one instruction.
- "Jump q," J q (where q is a numeral for an integer in $\mathbb{Z}$): All registers are left unchanged. The machine takes as its next instruction the qth instruction following this one in the program (if $q \geq 0$), or the $|q|$th instruction preceding this one (if $q < 0$). The machine halts if there is no such instruction in the program. An instruction of J 0 results in a loop, with the machine executing this one instruction over and over again.

Now suppose f is an n-place partial function on $\mathbb{N}$. Possibly there will be a program $\mathcal{P}$ such that if we start a register machine (having all the registers to which $\mathcal{P}$ refers) with $x_1, \ldots, x_n$ in registers $1, \ldots, n$ and 0 in the other registers, and we apply program $\mathcal{P}$, then the following conditions hold:

- If $f(x_1, \ldots, x_n)$ is defined, then the computation eventually terminates with $f(x_1, \ldots, x_n)$ in register 0. Furthermore, the computation terminates by seeking a $(p+1)$st instruction, where p is the length of $\mathcal{P}$.
- If $f(x_1, \ldots, x_n)$ is undefined, then the computation never terminates.

If there is such a program $\mathcal{P}$, we say that $\mathcal{P}$ *computes f*. (Notice when when we start a program running, there are three possibilities: (i) it might run forever; (ii) it might come to a "good" halt, by seeking the first instruction that isn't there; (iii) it might come to a "bad" halt, either by trying to jump back to somewhere before the start of the program or by trying to go forward to a nonexistent instruction other than the first such one. In our definition of "$\mathcal{P}$ computes f," we have elected to rule out bad halts. This will be a convenience in running programs end-to-end.)

For example, the addition function $x+y$ is computed by a register-machine program given in Chapter 1. Moreover, in Chapter 1, we saw a few basic subroutines:

CLEAR	r	(clear register r)
MOVE	from r to s	(register r is cleared)
COPY	from r to s using t	(register r is unchanged)

These subroutines have length 3, 7, and 15, respectively.

For a trivial example, the n-place constantly zero function is computed both by the empty program and by our three-line program to clear register 0:

$$
\begin{array}{lll}
\text{D} & 0 & \text{Try to decrement 0.} \\
\text{J} & 2 & \\
\text{J} & -2 & \text{Go back and repeat.} \\
& & \text{Halt.}
\end{array}
$$

(The arrows are to help us see what the program does.) Moreover, by adding some increment instructions, we can see that any total constant function is computed by some register program.

The one-place successor function $S(x) = x + 1$ is computed by the program that moves the contents of register 1 (call this number $[1]$) to register 0 and then increments the number:

$$
\begin{array}{lll}
\text{D} & 1 & \text{Take 1 from } [1]. \\
\text{J} & 3 & \text{Exit when zero.} \\
\text{I} & 0 & \text{Add 1 to } [0]. \\
\text{J} & -3 & \text{Repeat.} \\
\text{I} & 0 & \text{Add 1 to } [0].
\end{array}
$$

The projection function I_n^k is computed by the seven-line program that moves the contents of register n to register 0.

Next, we want to show that the class of partial functions computed by register-machine programs is closed under composition. That is, we want to know that whenever we have register-machine programs computing $f, g_1, \ldots, g_n$ and

$$h(\vec{x}) = f(g_1(\vec{x}), \ldots, g_n(\vec{x})),$$

then we can produce a program for h. This involves stringing several programs together. But care must be taken to be sure that one program does not trip over garbage left by an earlier program and does not erase data needed by a later program.

We know what it means for $\mathcal{P}$ to compute f: when we provide $\mathcal{P}$ with ideal conditions (the input in registers $1, \ldots, k$, the other registers empty), then $\mathcal{P}$ will return (in register 0) the function value, if it is defined.

But we need a program that is less fussy. What if conditions are not ideal? Suppose the input is in registers $r_1, \ldots, r_k$, where these are any k distinct numbers (not necessarily consecutive, and not necessarily in increasing order). And suppose we do not want to promise that the other registers are empty. Moreover, we want a program that does not erase or tamper with the contents of the first s registers, for some large number s – we want the information in those registers to be kept safe.

Here is what we want, formulated as a definition:

Definition: Suppose that f is a k-place partial function, Q is a program, $r_1, \ldots, r_k$ are distinct natural numbers, and s and t are natural numbers. Then, we say that Q *computes f from $r_1, \ldots, r_k$ to t preserving s* if whenever we start a register machine (with enough registers) with program Q and with $a_1, \ldots, a_k$ in registers $r_1, \ldots, r_k$, then, no matter what is in the other registers initially, we have the following end results:

- If $f(a_1, \ldots, a_k)$ is defined, then the computation eventually comes to a good halt (that is, it halts by seeking the first nonexistent instruction, the $(q + 1)$st instruction, where q is the length of Q), with the value $f(a_1, \ldots, a_k)$ in register t. Moreover, the first s registers, registers $0, 1, \ldots, s - 1$, contain the same numbers they held initially, except possibly for register t.
- If $f(a_1, \ldots, a_k)$ is undefined, then the computation never halts.

As a special case, we can say that if Q computes f from $1, \ldots, k$ to 0 preserving 0, then Q computes f (in the sense defined originally). The converse is not quite true because of the matter of whether registers are initially empty.

The following theorem says that we can have what the preceding definition asks for.

Lemma: *Assume that the program P computes the k-place partial function f. Let $r_1, \ldots, r_k$ be distinct natural numbers; let t and s be natural numbers. Then, we can find a program Q that computes f from $r_1, \ldots, r_k$ to t preserving s.*

Proof. Let M be the largest *address* in P (that is, the largest number such that some increment or decrement instruction in P addresses register M). Probably $M > k$; if not then increase M to be $k + 1$. Let j be the largest of the numbers $s, r_1, \ldots, r_k$. Here is program Q:

```
    COPY   r₁ to j + 1 using j + 2
    COPY   r₂ to j + 2 using j + 3

    COPY   rₖ to j + k using j + k + 1
    CLEAR  j
    CLEAR  j + k + 1
    CLEAR  j + k + 2
            . . .
    CLEAR  j + M
      P    relocated by j
    MOVE   from j to t   (if j ≠ t)
```

Here, "relocated by j" means that j is added to the address of all increment and decrement instructions. Thus, the relocated program operates on registers $j, j+1, \ldots, j+M$ exactly as P operated on registers $0, 1, \ldots, M$. Since P calculates f, the relocated program will leave $f(a_1, \ldots, a_k)$, if defined, in register j. The program Q then moves it to register t. Except for register t, the program leaves registers $0, 1, \ldots, j - 1$ unchanged. The following map illustrates how $\mathcal{Q}$ uses the registers:

register	0	untouched
register	1	untouched
	$\vdots$	
register	$j - 1$	untouched
register	j	output
register	$j + 1$	input
	$\vdots$	
register	$j + k$	input
register	$j + k + 1$	work space
	$\vdots$	
register	$j + M$	work space

The lemma will be a useful tool whenever we need to glue different programs together. As our first application of this lemma, we can show that the class of register-machine computable partial functions is closed under composition.

Theorem: *The class of register-machine computable partial functions is closed under composition. That is, whenever we have register-machine programs that compute partial functions $f, g_1, \ldots, g_n$ from which h is obtained by composition, then we can make a program that computes the partial function h.*

Proof. Suppose that the k-place partial function h is obtained by composition from f and $g_1, \ldots, g_n$:

$$h(\vec{x}) = f(g_1(\vec{x}), \ldots, g_n(\vec{x})).$$

Further suppose that we have programs that compute $f, g_1, \ldots, g_n$. We want to make a program that computes h. Here it is:

Calculate g_1 from $1, \ldots, k$ to $k + 1$, preserving $k + 1$.
Calculate g_2 from $1, \ldots, k$ to $k + 2$, preserving $k + 2$.
. . .
Calculate g_n from $1, \ldots, k$ to $k + n$, preserving $k + n$.
Calculate f from $k + 1, \ldots, k + n$ to 0, preserving 0.

Here, we rely on the lemma to provide the components of the program. Observe that for the program to halt, we need $g_1(\vec{x}), \ldots, g_n(\vec{x})$ to be defined, and we need f to be defined at $\langle g_1(\vec{x}), \ldots, g_n(\vec{x}) \rangle$. $\dashv$

For example, suppose that h is given by the equation

$$h(x, y, z) = f(g(z, y), 7, x)$$

and we have register programs for f and g. It follows from the preceding theorem that h is register-machine computable. We can write

$$h(x, y, z) = f(g\left(I_3^3(x, y, z), I_2^3(x, y, z)\right), K_7(x, y, z), I_1^3(x, y, z)$$

(where K_7 is a constant function) and apply the theorem twice, first to get $g(I_3^3(x, y, z), I_2^3(x, y, z))$ and then to get h. The moral of this example is that we can freely put the variables where we want and apply composition with projection functions to justify what we have done. In particular, if

$$h(x_1, x_2, \ldots, x_m) = f(\_\_\_, \_\_\_, \ldots, \_\_\_),$$

where each blank is filled by some x_i or some constant, then from a program for f we can obtain a program for h.

Still, this chapter has not yet produced a program for a single "interesting" function. For that, we need one more closure result: closure under primitive recursion. That is, we want to know that if h is obtained from f and g by primitive recursion

$$h(\vec{x}, 0) = f(\vec{x})$$
$$h(\vec{x}, y + 1) = g(h(\vec{x}, y), \vec{x}, y)$$

and we have register programs for f and g, then we can get a program for h. Or in the case where $\vec{x}$ is empty

$$h(0) = m$$
$$h(y + 1) = g(h(y), y)$$

(for some number m) and we have a program for g, then we want to know that we can get a program for h.

It will then follow that all primitive recursive functions (in particular, the ones in Chapter 2) are register-machine computable. And so finally, we will see that the class of register-machine computable functions includes much more than the simplistic examples we started with.

Theorem: *The class of register-machine computable partial functions is closed under primitive recursion.*

Proof. Assume that h is the partial $(n + 1)$-place function obtained by primitive recursion from the partial functions f and g:

$$h(\vec{x}, 0) = f(\vec{x})$$
$$h(\vec{x}, y + 1) = g(h(\vec{x}, y), \vec{x}, y)$$

Assume that we have register programs for f and g. We need to make a program for h.

The program will start with $x_1, \ldots, x_n, y$ in registers $1, 2, \ldots, n, n+1$. The program will put $h(\vec{x}, t)$ in register 0 first for $t = 0$, then for $t = 1$, and so forth up to $t = y$. The number t is kept in register $n + 2$, which initially contains 0. The following map illustrates this usage:

register	0	$h(\vec{x}, t)$
register	1	x_1
	$\vdots$	
register	n	x_n
register	$n + 1$	$y - t$
register	$n + 2$	t
register	$n + 3$	work space
	$\vdots$	

Here is the program:

Calculate f from $1, \ldots, n$ to 0, preserving $n + 3$

	D	$n + 1$	Begin loop.
	J	$\star + 3$	Halt when done.

Calculate g from $0, 1, \ldots, n, n + 2$ to 0, preserving $n + 3$

	I	$n + 2$	$t := t + 1$
	J	$-(\star + 3)$	Repeat loop.
			Halt.

Here $\star$ is the length of the program being used for g.

To see the correctness of this program, we establish the following:

Claim: Whenever we reach the D $n + 1$ instruction (at the beginning of the loop), after executing the loop k times,

- register 0 contains $h(\vec{x}, k)$,
- register $n + 1$ contains $y - k$,
- register $n + 2$ contains k.

The claim, stating "loop invariants," is proved by induction (as is usual, for programs with loops) on k.

For $k = 0$, when we reach D $n + 1$ without having executed the loop at all, register 0 contains $f(\vec{x})$, which *is* $h(\vec{x}, 0)$, register $n + 1$ is untouched so it still contains y, and register $n + 2$ is still 0.

Now for the inductive step. In the $(k + 1)$st pass through the loop, we decremented register $n + 1$ (by the inductive hypothesis it previously contained $y - k$, so now it is $y - (k + 1)$), we incremented register $n + 2$ (it previously contained k, so now it is $k + 1$), and in register 0, we put $g(h(\vec{x}, k), \vec{x}, k)$, which is indeed $h(\vec{x}, k + 1)$.

So by induction, the claim holds every time we start the loop. The program halts when we start the loop with 0 in register $n + 1$. At this point, we have done the loop y times (by the claim), and register 0 contains $h(\vec{x}, y)$, as desired.

The program is easily modified for the case where $\vec{x}$ is empty. We want to use the registers as follows:

register   0 $h(t)$
register   1 $y - t$
register   2 t
register   3   work space
$\vdots$

Here is the program:

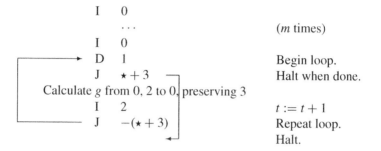

I	0	
$\cdots$		(m times)
I	0	
D	1	Begin loop.
J	$\star + 3$	Halt when done.
Calculate g from 0, 2 to 0, preserving 3		
I	2	$t := t + 1$
J	$-(\star + 3)$	Repeat loop.
		Halt.

The correctness argument continues to be applicable, with $n = 0$. $\dashv$

Gathering together the results thus far, we have the following conclusion.

Theorem: *Every primitive recursive function is register-machine computable.*

This theorem promises register-machine programs for all the primitive recursive functions from the previous chapter. But of course, we can do better:

Theorem: *Every general recursive partial function is register-machine computable.*

Proof. We need to add the μ-operator

$$h(\vec{x}) = \mu y[g(\vec{x}, y) = 0].$$

We use the obvious program:

Calculate g from 1, ..., n, 0 to $n + 1$, preserving $n + 2$		
D	$n + 1$	Are we there yet?
J	3	Halt.
I	0	$y := y + 1$
J	$-(\star + 3)$	Try again.
		Halt.

The program uses registers as follows:

register	0	y
register	1	x_1
$\vdots$		
register	n	x_n
register	$n+1$	$g(\vec{x}, y)$
register	$n+2$	work space
$\vdots$		

Of course, this program might never halt.   ⊣

This theorem gives half of a significant fact that formalizing the effective calcula-bility concept by means of general recursiveness and formalizing the effective calcu-lability concept by means of register machines lead to exactly the same class of partial functions. We will come to the other half soon. In particular, the theorem illustrates that register machines are capable of doing much more than might have been apparent initially from their very simple definition.

3.2   A Universal Program

In Chapter 1, it was argued that the "universal" partial function

$$\Phi(e, x) = \text{the result of applying the program coded by } e \text{ to input } x$$

should be a computable partial function. We now plan to verify this fact in the case of register-machine programs.

The first step is to make the phrase "program coded by e" meaningful, by adopting a particular coding.

First, to each instruction c, we will assign a number $\#c$ (called its *Gödel number*), as follows:

$$\#\text{I}\, r = [0, r]$$
$$\#\text{D}\, r = [1, r]$$
$$\#\text{J}\, q = [2, q] \text{ if } q \geq 0$$
$$\#\text{J}\, q = [3, |q|] \text{ if } q < 0$$

Here $[x, y] = 2^{x+1} \cdot 3^{y+1}$, as in the preceding chapter. In fact, we will utilize all of the sequence-coding machinery developed there. For example, $\#\text{I}\,0 = [0, 0] = 6$. And the instruction J -1 has the Gödel number $[3, 1] = 144$. Observe that applying $(\ )_0$ to the Gödel number of an instruction, we find out what type of instruction it is (increment, decrement, or jump). And by applying $(\ )_1$ to the Gödel number of an instruction, we find the address of the register or the size of the jump.

Next, to a program $c_0, \ldots, c_m$ (i.e., a finite string of instructions), we assign its Gödel number

$$[\#c_0, \ldots, \#c_m].$$

To the empty program, we assign the Gödel number 1. For example, the one-line program I 0 has the Gödel number $[6] = 2^7 = 128$. And the one-line program J -1 has the Gödel number $[[3, 1]] = 2^{145}$.

For example, what is $\Phi(128, 7)$? First, we decode $128 = [6] = [\#I\,0]$. Next, we apply this program to 7, obtaining an output of 1. We conclude that $\Phi(128, 7) = 1$. The function Φ is not total; $\Phi(2^{25}, 7)$ is undefined. We have not yet clarified what $\Phi(e, x)$ should be when e is not the Gödel number of any program; we will take care of that matter shortly.

Digression: We could have kept these numbers smaller by using a more efficient coding technique. A program is a string of symbols over an alphabet containing the symbols I, D, J, $-$, and whatever digits we use for the numerals specifying register numbers and jump lengths. As a program, J -1 is a word of length 3, in contrast to 2^{145} which, written out in base 10, takes 44 digits.

Suppose we use base-6 numerals to specify register numbers and jump lengths. For these numerals, we need the digits 0, 1, 2, 3, 4, and 5. So a program can be viewed as a word over the 10-letter alphabet

$$\{I, D, J, -, 0, 1, 2, 3, 4, 5\}.$$

One very natural way to code words over this alphabet is to use "decadic notation." That is, a word over a 10-letter alphabet *is* a numeral, and it names a number via a base-10 notation, albeit a notation somewhat different from the standard base-10 notation. For more about decadic notation, see the Appendix A3. But for our present purposes, efficiency of coding is not required.

In a similar spirit, we can next encode the contents of all the registers into a single number. Suppose that at some instant in time, the number z_i is in register i, for each i. This information we encode into the *memory number*

$$2^{z_0} \cdot 3^{z_1} \cdot \ldots \cdot p_i^{z_i} \cdot \ldots = \prod_i p_i^{z_i}.$$

This "infinite product" is finite because at any instant, only finitely many of the registers will be nonzero. (We deliberately did not use $z_i + 1$ in the exponent here.) Of course, any positive integer can be viewed as a memory number; we can take its prime factorization and read off the contents of each register. For example, a memory number of 243 tells us that register 1 contains 5 and the other registers are 0, because $243 = 3^5$.

With this coding in hand, we proceed to the second step, of constructing functions (which will turn out to be primitive recursive) for simulating the execution of a register-machine program.

One of these will be the "mem" function, which will specify how the memory number is changed by the execution of an instruction. That is, suppose that at some point, the register contents are given by the memory number m and then we execute the instruction with Gödel number c. We want mem(m, c) to be the new memory number, after the instruction is executed.

$$\text{mem}(m, c) = \begin{cases} m \cdot p_{(c)_1} & \text{if } (c)_0 = 0 \text{ and } c \neq 0 & \text{(increment)} \\ \lfloor m/p_{(c)_1} \rfloor & \text{if } (c)_0 = 1 \text{ and } p_{(c)_1} \mid m & \text{(decrement)} \\ m & \text{otherwise.} \end{cases}$$

For example, mem$(m, 6) = 2m$ for any number m. This is because 6 is the Gödel number of the instruction I 0, and executing this instruction increments the exponent of 2 in m's prime factorization. What is mem$(m, 12)$? 12 is the Gödel number of the instruction D 0. If m is even, then mem$(m, 12) = m/2$, which decrements the exponent of 2 in m's prime factorization. But if m is odd, then mem$(m, 12) = m$.

Observe that this equation for the mem function is expressed entirely within the language we have built up for primitive recursiveness. That is, simply from the *form* of the equation, we see that mem is a primitive recursive function. It is built up using definition by cases, the prime-counting function $p_n, \ldots$. And because we know that all primitive recursive functions are register-machine computable, we know we can make a register-machine program that computes mem.

Next suppose we have a program with Gödel number

$$e = [\#c_0, \ldots, \#c_m].$$

So lh $e = m + 1$, the number of instructions in the program. We want to think about the *location counter*, which keeps track of where we are in the program. If the location counter is 0, then we are about to execute c_0, the first instruction. More generally, if the location counter is k, then we are about to execute c_k, the $(k + 1)$st instruction. If the location counter is lh e, then the program has come to a good halt, by seeking the first nonexistent instruction. (A location counter greater than lh e would correspond to a bad halt, seeking a nonexistent instruction later than the first such.)

We want to define a function "loc" that gives the value of the location counter. That is, if the location counter is now k, then we want loc(k, m, e) to be the new value of the location counter after we execute c_k, the $(k + 1)$st instruction, where m is the memory number.

What does this function need to do? First, it needs to find $(e)_k = \#c_k$, the Gödel number of c_k. This number is a pair

$$(e)_k = [((e)_k)_0, ((e)_k)_1],$$

where $((e)_k)_0$ is 0, 1, 2, or 3. If $((e)_k)_0 = 1$, then we have a decrement instruction, and it will be necessary to check the memory number to see if the decrement is successful or not. Here is the function written out in full:

$$\text{loc}(k, m, e) = \begin{cases} k & \text{if } k \geq \text{lh } e & \text{(already halted)} \\ k + 2 & \text{if } ((e)_k)_0 = 1 \text{ and} \\ & p_{((e)_k)_1} \mid m \text{ and} & \text{(decrement)} \\ & k + 2 \leq \text{lh } e \\ \min(k + ((e)_k)_1, \text{lh } e) & \text{if } ((e)_k)_0 = 2 & \text{(jump forward)} \\ k \mathbin{\dot-} ((e)_k)_1 & \text{if } ((e)_k)_0 = 3 \text{ and} \\ & ((e)_k)_1 \leq k & \text{(jump back)} \\ \text{lh } e & \text{if } ((e)_k)_0 = 3 \text{ and} \\ & ((e)_k)_1 > k & \text{(bad jump)} \\ k + 1 & \text{otherwise} & \text{(default)} \end{cases}$$

For example, what is $\text{loc}(0, m, 128)$? Here $k = 0$, so we are at the very beginning of the program. And $e = 128$, which we need to decode. Since $e = 128 = 2^7$, we see that we have a one-line program (that is, $\text{lh}(e) = 1$). The one instruction has Gödel number $(e)_0 = 6$. We see that $((e)_0)_0 = 0$, so we have an increment instruction. And $((e)_0)_1 = 0$, so the register in question is register 0. (In other words, $6 = \#I0$.) In the above equation, it is the last line (the "default" line) that applies: $\text{loc}(0, m, 128) = k + 1 = 0 + 1 = 1$. The new value of the location counter is 1. This is just as it should be; having executed the $I0$ instruction, we push the location counter up one.

And next we come to $\text{loc}(1, 2m, 128)$. But the program has already halted. In the above equation, it is now the first line that is applicable. That is, because $1 = k \geq \text{lh } e = 1$, we obtain $\text{loc}(1, 2m, 128) = 1$.

The equation may be a bit unwieldy, but each piece of the equation does the natural thing: skip, jump forward, jump back, or go on to the next instruction, as the case may be. (If the program in question is trying to make a bad jump, either to a point before the beginning of the program or to a point later than the first nonexistent instruction, then the loc function generously tries to make the situation appear to be a little better than it really is.)

Again, from looking at the form of the equation, we realize that the loc function is primitive recursive. And being primitive recursive, it is therefore computable by a register program. (In particular, loc is a total function. Even if e is *not* the Gödel number of a program at all, $\text{loc}(k, m, e)$ will be defined. That is, if we put garbage into the loc function, then we get garbage out, but at least we get something out). And it is not hard to see that we always have $\text{loc}(k, m, e) \leq \max(k, \text{lh } e)$.

Now, we are ready to describe our universal program (that computes Φ). In computing $\Phi(e, x)$, initially e is in register 1 and x is in register 2, and the other registers are blank. The program will keep the location counter k in register 3 (initially 0), the memory number m in register 4, and the Gödel number $(e)_k$ of the next instruction in register 5.

register 0 number of remaining instructions
register 1 program e
register 2 input x
register 3 location k
register 4 memory m
register 5 instruction $(e)_k$
register 6 work space
$\vdots$

Here is our universal program, described in nine or ten lines:

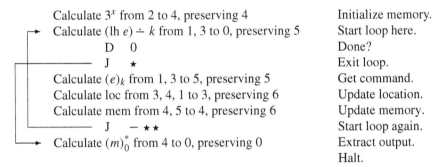

Calculate 3^x from 2 to 4, preserving 4   Initialize memory.
Calculate $(\text{lh } e) \doteq k$ from 1, 3 to 0, preserving 5   Start loop here.
          D   0   Done?
          J $\star$ Exit loop.
Calculate $(e)_k$ from 1, 3 to 5, preserving 5   Get command.
Calculate loc from 3, 4, 1 to 3, preserving 6   Update location.
Calculate mem from 4, 5 to 4, preserving 6   Update memory.
          J $- \star\star$ Start loop again.
Calculate $(m)_0^*$ from 4 to 0, preserving 0   Extract output.
                                                         Halt.

(Here $(m)_0^* = \mu t < m[2^{t+1} \nmid m]$, the exponent of 2 in the prime factorization of m.) This program, given x and e, decodes e to see what it says to do with x, and then does it (to paraphrase what was said on page 7).

First, consider the case where e is the Gödel number of a program $\mathcal{P}$ that computes a one-place partial function f. Then, the universal program will mimic the operation of $\mathcal{P}$ on input x. If $f(x)$ is defined, then the universal program will halt with $f(x)$ in register 0; if $f(x)$ is undefined, then the universal program will not halt. (What we have done is to make explicit just what "executing" a program $\mathcal{P}$ involves.)

Secondly, consider what happens if e is the Gödel number of a program $\mathcal{P}$ that does not compute any partial function. This can happen if $\mathcal{P}$ has some bad jumps (i.e., a jump to a nonexistent instruction other than the first such one). Again, the universal program will mimic the operation of $\mathcal{P}$ on input x. If a bad jump is encountered, the universal program will halt (by setting the location counter to lh e). But the universal program will come to a good halt (with output $(m)_0^*$, where m is the memory number at the time).

Thirdly, it may be that e is not the Gödel number of a program at all. The universal program will nonetheless start running. After all, both mem and loc are *total* functions. Maybe at the conclusion of some loop, the location counter will happen to equal lh e. Then, the universal program will halt (by seeking the first nonexistent instruction). And maybe that will never happen.

The point is that the universal program computes some partial two-place function Φ. Here, $\Phi(e, x)$ is whatever the universal program gives us, if and when it halts on input e and x.

Because Φ is a register-machine computable partial function, it follows that for any fixed e, the one-place function $x \mapsto \Phi(e, x)$ is register-machine computable (one program for it puts the constant e into register 2 and then runs the universal program). Call this function $[\![e]\!]$:

$$[\![e]\!](x) = \Phi(e, x).$$

That is, $[\![e]\!](x)$ is whatever this universal program produces (if anything), given the input $\langle e, x \rangle$.

Then, if e is the Gödel number of a program that computes a one-place partial function f, we can conclude that $[\![e]\!] = f$. And if e is some other number (either the number of a program that does not compute a partial function or perhaps not the Gödel number of any program at all), then we can say at least that $[\![e]\!]$ is *some* register-machine computable partial function.

Whenever e is a number for which $[\![e]\!]$ is the function f, we will say that e is an *index* of f. Thus, the indices of computable partial function f include the Gödel numbers of programs that compute f and also include any other numbers for which $[\![e]\!]$ just happens to be the function f (that is, $\Phi(e, x) = f(x)$ for all x).

In summary, we now have the following result.

Enumeration theorem:
  (i) Φ *is a register-machine computable two-place partial function.*
  (ii) *For each number e, $[\![e]\!]$ is a register-machine computable one-place partial function.*
  (iii) *Each one-place register-machine computable partial function is $[\![e]\!]$ for some number e.*

Thus,

  $[\![0]\!]$, $[\![1]\!]$, $[\![2]\!]$, $\ldots$

is a complete list (with repetitions) of all the one-place register-machine computable partial functions, and only those.

It is straightforward to generalize these ideas from one-place partial functions to n-place partial functions. Our universal programs start with $e, x_1, x_2, \ldots, x_n$ in registers $1, 2, \ldots, n, n + 1$ and with 0 in the other registers. The program will keep the location counter k in register $n + 2$, the memory number m in register $n + 3$, and the Gödel number $(e)_k$ of the next instruction in register $n + 4$. Here it is:

> Calculate $3^{x_1} 5^{x_2} \cdots p_n^{x_n}$ from $2, 3, \ldots, n + 1$ to $n + 3$, preserving $n + 3$
> Calculate $(\mathrm{lh}\ e) \dot- k$ from $1, n + 2$ to 0, preserving $n + 4$
> D   0
> J $\star$
> Calculate $(e)_k$ from $n, n + 2$ to $n + 3$, preserving $n + 3$
> Calculate loc from $n + 2, n + 3, n + 1$ to $n + 2$, preserving $n + 5$
> Calculate mem from $n + 3, n + 4$ to $n + 3$, preserving $n + 5$
> J $- \star \star$
> Calculate $(m)_0^*$ from $n + 3$ to 0, preserving 0

This program computes an $(n + 1)$-place partial function $\Phi^{(n)}$. For each fixed e, define the n-place partial function $[\![e]\!]^{(n)}$ by the equation

$$[\![e]\!]^{(n)}(x_1, x_2, \ldots x_n) = \Phi^{(n)}(e, x_1, x_2, \ldots, x_n).$$

Enumeration theorem:

(i) $\Phi^{(n)}$ *is a register-machine computable $(n + 1)$-place partial function.*

(ii) *For each number e, $[\![e]\!]^{(n)}$ is a register-machine computable n-place partial function.*

(iii) *Each n-place register-machine computable partial function is $[\![e]\!]^{(n)}$ for some number e.*

Thus,

$$[\![0]\!]^{(n)}, \ [\![1]\!]^{(n)}, \ [\![2]\!]^{(n)}, \ \ldots$$

is a complete list (with repetitions) of all the n-place register-machine computable partial functions, and only those.

One significant benefit of having universal functions is that we can apply diagonalization to them. The diagonal function

$$d(x) = [\![x]\!](x) + 1$$

is a register-machine computable partial function (being obtained from Φ, I_1^1, and S by composition). And so $d = [\![e]\!]$ for some number e. What can we say about $d(e)$? We have the equation

$$d(e) = [\![e]\!](e) + 1 = d(e) + 1,$$

but "$=$" means that either both sides are undefined or both sides are defined and are equal. We can conclude the $d(e)$ must be undefined, lest $0 = 1$.

(Another diagonal function

$$\hat{d}(x) = 1 \mathbin{\dot{-}} [\![x]\!](x)$$

would serve just as well here. And $\hat{d}$ has the added advantage of being bounded by 1.)

By contrast, suppose we attempt to change d into a total function:

$$D(x) = \begin{cases} [\![x]\!](x) + 1 & \text{if this is defined} \\ 0 & \text{otherwise} \end{cases}$$

Then, D is *not* register-machine computable. We cannot possibly have $D = [\![e]\!]$ because either $[\![e]\!](e)$ is defined and $D(e)$ is larger by 1 or else $[\![e]\!](e)$ is undefined and $D(e) = 0$. In fact, the same argument yields a slightly stronger statement.

Theorem:

(a) *There is no total register-machine computable function that extends the diagonal function $d(x) = [\![x]\!](x) + 1$.*

(b) *There is no total register-machine computable function that extends the diagonal function $\hat{d}(x) = 1 \mathbin{\dot{-}} [\![x]\!](x)$.*

Let K be the domain of the diagonal function:

$$K = \{x \mid [\![x]\!](x) \downarrow\}.$$

The semicharacteristic function of K

$$c_K(x) = \begin{cases} 1 & \text{if } x \in K \\ \uparrow & \text{if } x \notin K \end{cases}$$

is a register-machine computable partial function. To compute $c_K(x)$, we first try to compute $[\![x]\!](x)$; if and when we succeed, we give output 1. Or in equation form, $c_K(x) = 1 + 0 \cdot [\![x]\!](x)$.

But now consider the full characteristic function of K:

$$C_K(x) = \begin{cases} 1 & \text{if } x \in K \\ 0 & \text{if } x \notin K. \end{cases}$$

Theorem: C_K *is not register-machine computable.*

Proof. Suppose that, to the contrary, C_K was register-machine computable. Then, the above diagonal function D would be computed by the following program:

Calculate $C_K(x)$ from 1 to 0, preserving 2	Decide if $x \in K$.
D 0	Yes?
J $\star + 1$	Halt.
Calaculate $d(x)$ from 1 to 0, preserving 2	
	Halt.

where $\star$ is the length of the program used for d. $\dashv$

Unsolvability of the halting problem: *The total function*

$$H(x, y) = \begin{cases} 1 & \text{if } [\![x]\!](y) \downarrow \\ 0 & \text{if } [\![x]\!](y) \uparrow \end{cases}$$

is not register-machine computable.

Proof. $C_K(x) = H(x, x)$. $\dashv$

In terms of binary relations, this result says that the characteristic function of

$$\{\langle x, y \rangle \mid [\![x]\!](y) \downarrow\}$$

(which is the domain of Φ) is not register-machine computable. Its semicharacteristic function *is* register-machine computable, as in the case of K.

What we are doing here is retracing some of the material that appeared in Chapter 1 in informal terms, but now formalized by using register machines as our model of computability, and by exploiting the development in Chapter 2 of general recursive functions.

We can extract even more from these ideas, if we add a "clock." That is, instead of looking at $\Phi(e, x)$, we will add a third variable for time t and determine, for a triple $\langle e, x, t \rangle$, where the calculation of the program with Gödel number e and input x stands after t steps. More specifically, we want to determine both the location counter and the memory number after t steps.

The pair

[location counter, memory number]

gives us a "snapshot" showing the status of the calculation. So what we want is a "snap" function such that snap(e, x, t) gives the snapshot after t steps.

For a start, what is snap$(e, x, 0)$? No steps have been executed, so nothing has happened yet. The location counter is 0, and the memory number is 3^x:

$$\text{snap}(e, x, 0) = [0, 3^x],$$

which happens to be $2 \cdot 3^{(3^x+1)}$. For example, a snapshot of 162 tells us that the location counter is 0, register 1 contains 1, and the other registers are 0. (This holds because $162 = 2 \cdot 81 = 2 \cdot 3^{3+1} = [0, 3]$.)

Now suppose we know, for some number t of steps, the snapshot

$$\text{snap}(e, x, t) = [k, m].$$

What comes next? The next instruction to execute is $(e)_k$, if $k < \text{lh } e$. The memory number will change from m to mem$(m, (e)_k)$. The location counter will change from k to loc(k, m, e). Putting these pieces together gives us the equation:

$$\text{snap}(e, x, t + 1) = [\text{loc}(k, m, e), \text{mem}(m, (e)_k)].$$

Or noting that $k = (\text{snap}(e, x, t))_0$ and $m = (\text{snap}(e, x, t))_1$ yields the equation:

$$\text{snap}(e, x, t + 1) =$$
$$[\text{loc}((\text{snap}(e, x, t))_0, (\text{snap}(e, x, t))_1, e), \text{mem}((\text{snap}(e, x, t))_1, (e)_{(\text{snap}(e,x,t))_0})].$$

So we now have a pair of recursion equations for snap:

$$\text{snap}(e, x, 0) = [0, 3^x]$$
$$\text{snap}(e, x, t + 1) =$$
$$[\text{loc}((\text{snap}(e, x, t))_0, (\text{snap}(e, x, t))_1, e), \text{mem}((\text{snap}(e, x, t))_1, (e)_{(\text{snap}(e,x,t))_0})].$$

Moreover, the recursion equations use only known primitive recursive pieces. We conclude that the snap function is primitive recursive (and hence register-machine computable).

Observe that snap is a *total* function. Even if e is the Gödel number of a weird program, or even if e is not the Gödel number of a program at all, the quantity snap(e, x, t)

is defined for all x and t. Compare this with the universal program for $\Phi(e, x)$. In both cases, we are starting with a memory number of 3^x and a location counter of 0 and then applying the functions mem and loc over and over. But for snap(e, x, t), we get to stop after t steps.

Of course, we are particularly interested in the case where e is indeed the Gödel number of a program $\mathcal{P}$ that computes a partial function f. In this case, whenever $x \in \text{dom} f$, then sooner or later we will reach a snapshot where the location counter says the computation has halted:

$$(\text{snap}(e, x, t))_0 \geq \text{lh } e.$$

We might think of this situation as a cause for celebration. Accordingly, we define the ternary relation T

$$T = \{\langle e, x, t \rangle \mid (\text{snap}(e, x, t))_0 \geq \text{lh } e\},$$

or in other words,

$$T(e, x, t) \iff (\text{snap}(e, x, t))_0 \geq \text{lh } e$$
$$\iff [\![e]\!](x)\downarrow \text{ in } \leq t \text{ steps.}$$

We observe that the relation T is primitive recursive. And the partial function

$$\langle e, x \rangle \mapsto \mu t\, T(e, x, t)$$
$$\mapsto \mu t\, ([\![e]\!](x)\downarrow \text{ in } \leq t \text{ steps})$$

measures the running time of e at x (where the running time is undefined when the computation goes on forever).

In general, we cannot put an upper bound on the μ-operator here. And by the unsolvability of the halting problem, we do not in general know when the search for t will succeed and when it will go on forever. Nonetheless, we can at least assert that the running-time function is a *general* recursive *partial* function because it is obtained by applying search to a primitive recursive relation. (It is certainly not a total function.)

The "terminal snapshot" is snap$(e, x, \mu t\, T(e, x, t))$, if this is defined at all. (We can think of this as $\lim_{t\to\infty} \text{snap}(e, x, t)$, a limit that might or might not exist. Once we reach the terminal snapshot, if we do, then the snap function stops changing – the functions mem and loc have been constructed in such a way as to make sure of this.) From the terminal snapshot, we can apply $(\ )_1$ to obtain the terminal memory number. And then, we can apply to that quantity the function $(\ )_0^*$ to obtain the contents of register 0 at termination. And that is the output of the calculation. Thus, we obtain the following conclusion:

Normal form theorem: *For any x and e,*

$$[\![e]\!](x) = \Phi(e, x) = ((\text{snap}(e, x, \mu t\, T(e, x, t)))_1)_0^*$$
$$= ((\text{snap}(e, x, \mu t[(\text{snap}(e, x, t))_0 \geq \text{lh } e]))_1)_0^*$$

where as usual "=" means that either both sides are undefined or else both sides are defined and equal.

In other words, we can break down the calculation of $[\![e]\!](x)$ into four steps:

- Find $t_{halt} = \mu t\, T(e, x, t)$.
- Find the terminal snapshot $snap_{halt} = snap(e, x, t_{halt})$.
- Find the memory number $m_{halt} = (snap_{halt})_1$.
- Find the output value $(m_{halt})_0^*$.

The first step uses a general recursive function; the other steps use primitive recursive functions.

It is straightforward to extend these ideas to functions of more than one variable: $snap^{(2)}(e, x_1, x_2, t)$ is obtained by starting with

$$snap^{(2)}(e, x_1, x_2, 0) = [0, 3^{x_1} \cdot 5^{x_2}]$$

and proceeding as before. We define the $(n + 2)$-ary relation $T^{(n)}$

$$T^{(n)}(e, \vec{x}, t) \iff (snap^{(n)}(e, \vec{x}, t))_0 \geq lh\, e$$

$$\iff [\![e]\!]^{(n)}(\vec{x}) \downarrow \text{ in } \leq t \text{ steps.}$$

Again we observe that the relation $T^{(n)}$ is primitive recursive.

Normal form theorem: *For any n, e, and $\vec{x}$,*

$$[\![e]\!]^{(n)}(\vec{x}) = \Phi^{(n)}(e, \vec{x}) = ((snap^{(n)}(e, \vec{x}, \mu t\, T^{(n)}(e, \vec{x}, t)))_1)_0^*$$

$$= ((snap^{(n)}(\vec{x}, e, \mu t[(snap^{(n)}(e, \vec{x}, t))_0 \geq lh\, e]))_1)_0^*.$$

Looking at the right-hand side in this equation, we observe that it defines a general recursive partial function. In fact, everything on the right side is primitive recursive, except for the single application of the μ-operator. Hence, we have the following:

Corollary: *Every register-machine computable partial function is general recursive.*

Putting this corollary together with an earlier theorem, we conclude that the class of register-machine computable partial functions is exactly the same as the class of general recursive functions. This result is a welcome byproduct of our analysis of register-machine computations.

The methods used here to obtain the equivalence of general recursiveness to register-machine computability are adaptable to obtaining equivalence between other formalizations of effective calculability that were described in Chapter 1. For example, to show that Turing machines compute *at least* as many functions as the other approaches give us, it suffices to show first that Turing machines can compute the zero, successor, and projection functions; and secondly that the class of Turing-computable partial functions is closed under composition, primitive recursion, and search. In the other direction, to show that Turing machines compute *at most* the functions given by the other approaches, we can code the Turing machines in a suitable way, construct a universal Turing machine, and prove a normal form theorem. There are textbooks that do exactly that.

As another consequence of the normal form theorem, we can represent the domain of the partial function $[\![e]\!]$ in the following form:

$$x \in \text{dom } [\![e]\!] \iff \exists t[(\text{snap}(e, x, t))_0 \geq \text{lh } e]$$
$$\iff \exists t\, T(e, x, t)$$
$$\vec{x} \in \text{dom } [\![e]\!]^{(n)} \iff \exists t[(\text{snap}^{(n)}(e, \vec{x}, t))_0 \geq \text{lh } e]$$
$$\iff \exists t\, T^{(n)}(e, \vec{x}, t)$$

On the right, we have a primitive recursive relation, prefixed by an (unbounded) "$\exists t$" quantifier.

Digression: The relation T (or more generally, $T^{(n)}$) that we have constructed is closely related to what is generally called "the Kleene T-predicate." There are technical differences, however.

Exercises

1. Give a register-machine program that computes the function

$$Z(x) = \begin{cases} 1 & \text{if } x = 0 \\ 0 & \text{if } x > 0. \end{cases}$$

2. (a) Show that the set of Gödel numbers of instructions is a primitive recursive set.
   (b) Show that the set of Gödel numbers of programs is a primitive recursive set.
3. Determine $[\![0]\!](x)$ for all values of x for which this is defined. (Note that 0 is not the Gödel number of any program.)
4. We know that the one-place function that is constantly equal to k is computable. Show that there is a primitive recursive function f such that $[\![f(k)]\!]$ is that function. That is, we need the equation $[\![f(k)]\!](x) = k$ to hold for all k and x.
5. Show that the partial function

$$\text{time}(e, x) = \mu t\, T(e, x, t)$$

is not bounded by any total computable function. That is, show that there is no total computable function F with the property that

$$\text{time}(e, x) \leq F(e, x)$$

whenever the left side is defined.
6. Assume that h is a primitive recursive function and e is the Gödel number of a program such that

$$[\![e]\!](x) \downarrow \text{ in } \leq h(x) \text{ steps}$$

for all x. Show that the function $[\![e]\!]$ is primitive recursive. (That is, a program that runs in primitive recursive time always computes a primitive recursive function.)
7. Explain why, for a calculation that eventually halts, all the snapshots that arise in the course of the calculation must be distinct, prior to the terminal snapshot.

3.3 Register Machines Over Words[1]

The inputs to effective procedures are not really numbers, but numerals – strings of symbols. For example, the input to a Turing machine consists of a string of symbols written on consecutive squares of its tape. The register machines we have been considering up to now can be thought of as working with "base-1" numerals, where the numeral for 7 is the string

   |||||||

of seven tally marks. In base-1 notation, the increment and decrement commands are the natural ones to use.

But suppose that we wanted our machines to work with binary numerals. In this case, each register would contain some string (possibly empty) of 0's and 1's. Now the increment command seems less natural; changing 111 to 1000 is a big change, in some sense – every symbol changes.

We want to extend the register-machine concept to the situation where each register contains a string (possibly empty) of symbols from a q-letter alphabet

$$\Sigma = \{a_1, \ldots, a_q\}.$$

Actually, the alphabet is an *ordered* set,

$$\Sigma = \langle a_1, \ldots, a_q \rangle$$

because alphabetical order will matter.

For the machines we have been considering up to now, $q = 1$ and $\Sigma = \langle | \rangle$. For binary notation, we would use $q = 2$ and $\Sigma = \langle 0, 1 \rangle$.

To simplify the exposition, we will fix a particular size of alphabet, namely $q = 3$, where $\Sigma = \langle a, b, c \rangle$. But it will be clear how to adjust the concepts to larger or smaller values of q.

The commands, as before, are of three types: increment, decrement, and jump:

- "Increment register r by a," $I_a\ r$ (where r is a numeral for a natural number): The effect of this instruction is to prefix the letter a to the (left) end of the word in register r. If the register was previously empty, then it will now contain the one-letter word a. The machine then proceeds to the next instruction in the program (if any).
- "Increment register r by b," $I_b\ r$ (where r is a numeral for a natural number): The effect of this instruction is the same, but it prefixes the letter b to the (left) end of the word in register r.
- "Increment register r by c," $I_c\ r$ (where r is a numeral for a natural number): This instruction prefixes the letter c.
- "Decrement register r," $D\ r$ (where r is a numeral for a natural number): The effect of this instruction depends on the contents of register r. If the word in register r is empty, the machine simply proceeds to the next instruction, without changing the contents of the register. But if the word is nonempty, then the last (rightmost) letter is deleted. And what the machine does next depends on that deleted letter.

[1] This material will be needed in Chapter 7.

- If the deleted letter was a, then the machine skips one instruction and goes to the next one after that (if any).
- If the deleted letter was b, then the machine skips two instructions and goes to the next one after that (if any).
- If the deleted letter was c, then the machine skips three instructions and goes to the next one after that (if any).

In summary, the machines tries to delete the last letter in register r, and if it is successful, then it skips the appropriate number of instructions.

- "Jump n," J n (where n is a numeral for an integer in $\mathbb{Z}$): All registers are left unchanged. The machine takes as its next instruction the nth instruction following this one in the program (if $n \geq 0$), or the $|n|$th instruction preceding this one if ($n < 0$). The machine halts if there is no such instruction in the program. An instruction of J 0 results in a loop, with the machine executing this one instruction over and over again.

Observe that in the case of a *one*-letter alphabet, the preceding list of commands is the same list we considered before.

Example: The five-line program (call it CLEAR 3)

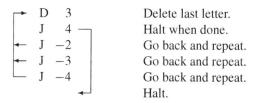

D	3	Delete last letter.
J	4	Halt when done.
J	−2	Go back and repeat.
J	−3	Go back and repeat.
J	−4	Go back and repeat.
		Halt.

will clear whatever is in register 3.

Example: The following program will take the word in register 1 and prefix it to whatever was in register 2. So if register 2 was empty, this program will simply move the word from register 1 to register 2. In any case, the concatenation of the words in registers 1 and 2 will be placed in register 2. At the end, register 1 will be empty.

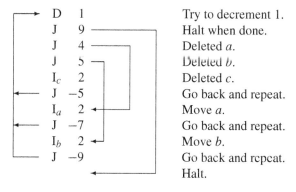

D	1	Try to decrement 1.
J	9	Halt when done.
J	4	Deleted a.
J	5	Deleted b.
I$_c$	2	Deleted c.
J	−5	Go back and repeat.
I$_a$	2	Move a.
J	−7	Go back and repeat.
I$_b$	2	Move b.
J	−9	Go back and repeat.
		Halt.

This program consists of 10 instructions. Call it PREFIX 1 to 2.

We can use this program to do a "right increment," that is, to *append* a letter to a given word. Assume we know that register 7 is empty. Then, the following program will append the letter b to the right end of whatever word is in register 0:

$$\text{I}_b \quad 7$$
$$\text{PREFIX 0 to} \quad 7$$
$$\text{PREFIX 7 to} \quad 0$$

Call this APPEND b to 0. It consists of 21 instructions. In using this program, we need to know of an empty register (in this example, register 7).

Now suppose we think of the words over our alphabet as being *numerals*, that is, as naming numbers. Because we have a three-letter alphabet, we treat words as base-3 numerals. More specifically, we use *triadic* notation, where the letters a, b, and c name 1, 2, and 3, respectively:

$$v(a) = 1, \quad v(b) = 2, \quad v(c) = 3.$$

Then, the general rule is that a k-letter word

$$s_k s_{k-1} \cdots s_1$$

names $v(s_k)3^{k-1} + v(s_{k-1})3^{k-2} + \cdots + v(s_2)3 + v(s_1)$, and the empty word λ names 0. Then, we obtain a one-to-one correspondence between the set of all words and the set of natural numbers. (See Appendix A3 for a discussion of these numerals.) Here is a list of the first few numerals:

Numeral	Number
λ	0
a	1
b	2
c	3
aa	4
ab	5
ac	6
ba	7
bb	8
bc	9
ca	10
cb	11
cc	12
aaa	13
aab	14
$\cdots$	$\cdots$

How do we add one in triadic notation? We look at the rightmost letter in the numeral. If that letter is a or b, then we simply increase it to the next letter. But if the rightmost letter is c, then we replace it by a (this lowers the number by 2), and we carry 1 to the left (which raises the number by 3). Here are some examples:

$$bca + a = bcb, \quad ccc + a = aaaa, \quad acc + a = baa.$$

Let's make a register-machine program to do this. That is, we want a program that computes the successor function, in triadic notation.

Assume the given word (the given numeral) is in register 1, and that register 0 is initially empty.

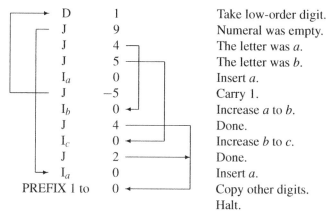

D	1	Take low-order digit.
J	9	Numeral was empty.
J	4	The letter was a.
J	5	The letter was b.
I_a	0	Insert a.
J	−5	Carry 1.
I_b	0	Increase a to b.
J	4	Done.
I_c	0	Increase b to c.
J	2	Done.
I_a	0	Insert a.
PREFIX 1 to	0	Copy other digits.
		Halt.

This program leaves the output in register 0, with register 1 empty. Call it ADD1 from 1 to 0.

Subtracting one (that is, computing the predecessor function) is very similar, except that instead of carrying to the left, we borrow from the left. And of course, we cannot subtract from λ, the numeral for 0.

Here are some examples:

$$bcb - a = bca, \quad baa - a = acc, \quad aaaa - a = ccc.$$

In general, we look at the rightmost digit. If it is b or c, then we simply lower it to a or b, respectively, and we are done. But if it is a, then we replace it by c and borrow one from the left.

Assume the given word (the given numeral) is in register 1, and that register 0 is initially empty.

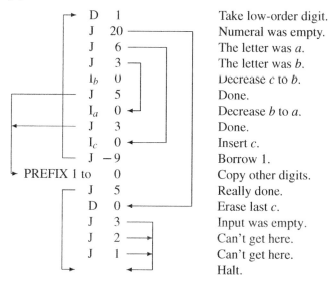

D	1	Take low-order digit.
J	20	Numeral was empty.
J	6	The letter was a.
J	3	The letter was b.
I_b	0	Decrease c to b.
J	5	Done.
I_a	0	Decrease b to a.
J	3	Done.
I_c	0	Insert c.
J	−9	Borrow 1.
PREFIX 1 to	0	Copy other digits.
J	5	Really done.
D	0	Erase last c.
J	3	Input was empty.
J	2	Can't get here.
J	1	Can't get here.
		Halt.

This program leaves the output in register 0, with register 1 empty. Call it SUB1 from 1 to 0. If applied to the empty word, the output is also empty.

Theorem: *Every n-place general recursive partial function f is register-machine computable in the following sense. There is a program $\mathcal{P}$ such that if we start a register machine with the triadic numerals for $x_1, \ldots, x_n$ in registers $1, \ldots, n$ and λ in the other registers and we apply program $\mathcal{P}$, then the following conditions hold:*

- *If $f(x_1, \ldots, x_n)$ is defined, then the computation eventually terminates with the triadic numeral for $f(x_1, \ldots, x_n)$ in register 0. Furthermore, the computation terminates by seeking a $(p+1)$st instruction, where p is the length of $\mathcal{P}$.*
- *If $f(x_1, \ldots, x_n)$ is undefined, then the computation never terminates.*

The proof is much as in the base-1 case, *mutatis mutandis*. The zero functions are computed by the empty program and by many others. The successor function is computed by ADD1 from 1 to 0. The projection function I_n^k is computed by PREFIX n to 0.

Closure under composition is a matter of good organization and careful bookkeeping. For closure under primitive recursion, we use both the SUB1 program and the ADD1 program. Closure under the μ-operator uses the ADD1 program.

Tools for writing and evaluating register-machine programs over a two-letter alphabet have been made available on the Web by Lawrence Moss; see http://www.indiana.edu/~iulg/trm. The alphabet used there is {1, #}.

Exercises

8. Modify the program PREFIX 1 to 2 for a two-letter alphabet $\Sigma = \langle a, b \rangle$.
9. Give a program that takes the first (i.e., leftmost) symbol (if any) from the word in register 2 and puts it into register 1 (assumed to be initially empty). At the end, register 2 should be empty.
10. Give a program (call it REVERSE 1 to 0) that takes the input word in register 1 and copies it into register 0, backwards. That is, the letters in the output word must be the same as the letters in the input word, but in the opposite order.
11. Modify the program REVERSE 1 to 0 from the previous exercise for a two-letter alphabet $\Sigma = \langle a, b \rangle$.
12. Give a program that takes the last (i.e., rightmost) letter from the word in register 1 and prefixes it to the left of the word in register 0. But at the end, the word in register 1 is to be unchanged.
13. Modify the program ADD1 from 1 to 0 for a two-letter alphabet $\Sigma = \langle a, b \rangle$.
14. Modify the program SUB1 from 1 to 0 for a two-letter alphabet $\Sigma = \langle a, b \rangle$.

3.4   Binary Arithmetic

Addition and multiplication are primitive recursive functions. So by a recent theorem, there are programs to compute them (in triadic notation). But the proof to that theorem yields very slow programs. The program for addition would calculate $x + y$ by going through a loop y times, adding 1 to x each time it executes the loop. If y is a huge number, this is going to take a huge amount of time; it is going to take an amount of time proportional to y.

The way we learned to add in second grade is a great deal faster. We start by adding the low-order digits of x and y. This gives us the low-order digit of the sum, and tells us whether or not we need to carry a digit. Then, we keep going until we are done. The amount of time will be roughly proportional to the *length* of the numeral for y.

Let's look at this a little more carefully. But changing the scene, we take the two-letter alphabet $\Sigma = \langle 0, 1 \rangle$, and we use standard binary numerals (not dyadic numerals). So for addition, we need a binary adder. If y is a 100-bit number (i.e., $|y| = 100$, where $|y|$ is the number of bits in the binary numeral for the number y), we expect to go through the program's loop 100 times. But a 100-bit number is at the very least 2^{99}, and we definitely do not want to go through the program's loop 2^{99} times.

In outline, we know how a binary adder works. Initially, x is in register 1 (in binary), y is in register 2 (in binary), and the other registers contain the empty word λ. The $(k + 1)$st time through the main loop (initially $k = 0$), register 0 will contain the k low-order bits of the sum, registers 1 and 2 will contain x and y except for their k low-order bits, and register 3 will contain either 1 (to show a carry bit) or λ (to show there is no carry bit).

How long will this program take? Each time through the loop, we shorten the words in registers 1 and 2 by one bit. So the number of times we go through the loop is bounded by $\min(|x|, |y|)$, the length of the shorter input numeral. Once we exit the main loop, the leftover bits from the longer input numeral need to be prefixed to the sum numeral. So the number of steps to obtain the sum will be

$$\text{constant} \times \max(|x|, |y|),$$

where the constant depends on the program, but will be less than the length of the program.

Now what about multiplication? Again, it would be much too slow to compute xy by adding x to itself y times. In the third grade, we all learned a much faster algorithm. This algorithm involves going through a certain loop a number of times equal to $|y|$, the *length* of y. Each time through the loop, we either do nothing (if the bit in y is 0) or add x – suitably shifted – to the sum being accumulated (if the bit in y is 1).

We could code this algorithm into a suitable program. But instead, let's leap to the real point: the program will produce xy (in binary) in a number of steps that is bounded by

$$\text{constant} \times |y| \times \max(|x|, |y|)$$

for some constant.

4 Recursive Enumerability

First of all, let's summarize some of the results of the preceding chapters and establish the terminology that will be used henceforth.

We have seen that the class of general recursive partial functions is exactly the same as the class of register-machine computable partial functions. The fact that two such different approaches yield the same class of functions is evidence that we have here a "natural" class. The members of this class will be called *computable partial functions* (or *recursive partial functions* – the two names are both in common use). The adjective "partial" covers both the total and nontotal functions; it can be omitted in cases where we know that the function is total. Church's thesis is the assertion that the concept of being a computable partial function is the correct formalization of the informal idea of being an effectively calculable partial function.

The class of computable partial functions includes all of the primitive recursive functions. Moreover, the class is closed under composition, the μ-operator, and definition by cases, among other things.

We define a relation R on $\mathbb{N}$ to be a *computable relation* (or a *recursive relation*) if its characteristic function (which is always total) is a computable function. The class of computable relations includes all of the primitive recursive relations. Moreover, the class is closed under unions, intersections, complements, and bounded quantification. Church's thesis tells us that the concept of being a computable relation corresponds to the informal idea of being a decidable relation.

Moreover, we have found the following basic result.

Enumeration theorem: *For each positive n, there is an $(n + 1)$-place computable partial function $\Phi^{(n)}$ with the property that for every n-place computable partial function f, there exists a number e such that*

$$\Phi^{(n)}(e, \vec{x}) = f(\vec{x})$$

for all n-tuples $\vec{x}$, where "=" has the usual meaning – either both sides are undefined or they are both defined and are the same.

Consequently, we can let $[\![e]\!]^{(n)}$ be the *n*-place partial function defined by the equation

$$[\![e]\!]^{(n)}(\vec{x}) = \Phi^{(n)}(e, \vec{x})$$

and obtain a complete enumeration (with repetitions)

$$[\![0]\!]^{(n)}, \quad [\![1]\!]^{(n)}, \quad [\![2]\!]^{(n)}, \quad \ldots$$

of all of the n-place computable partial functions. When $n = 1$, we can omit the super-script.

The enumeration theorem lets us define the "diagonal" partial function

$$d(x) = [\![x]\!](x) + 1.$$

This is a one-place computable partial function; its domain is the set

$$K = \{x \mid [\![x]\!](x) \downarrow\}.$$

Because the diagonal function d is a computable partial function, we know that $d = [\![e]\!]$ for some index e. That is, $[\![x]\!](x) + 1 = d(x) = [\![e]\!](x)$ for all x. In particular, taking $x = e$, we have $[\![e]\!](e) + 1 = d(e) = [\![e]\!](e)$. This looks like bad news; we *almost* have proved that $1 = 0$. But remember that "=" means that either both sides are defined and equal, or both sides are undefined. Here it must be the latter. A modification of the argument gives us the following result.

Theorem: *The diagonal function d has no extension to a computable total function. (That is, there is no computable total function f with the property that whenever $d(x)$ is defined then $f(x) = d(x)$.)*

Proof. Suppose that $[\![e]\!]$ is a computable partial function extending d. (Maybe it even is d.) We will show that $e \notin \mathrm{dom}\, [\![e]\!]$, thereby showing that $[\![e]\!]$ is not total.

If to the contrary $e \in \mathrm{dom}\, [\![e]\!]$, then $e \in K$ and $d(e) \downarrow$. But then, we have

$$[\![e]\!](e) = d(e) \quad \text{because } [\![e]\!] \text{ extends } d$$
$$= [\![e]\!](e) + 1 \quad \text{by the definition of } d$$

and these are defined. Hence $0 = 1$, a contradiction. $\dashv$

The same argument would apply to the function $\hat{d}(x) = 1 \dot- [\![x]\!](x)$.

Corollary: *The set K is not a computable set.*

Proof. The function

$$f(x) = \begin{cases} d(x) & \text{if } x \in K \\ 0 & \text{if } x \notin K \end{cases}$$

is a total extension of d and therefore is not a computable function. But if K were a computable set, then f would have been a computable function (by definition by cases). $\dashv$

Unsolvability of the halting problem: *The relation*

$$H = \{\langle x, y \rangle \mid [\![x]\!](y) \downarrow\}$$

is not a computable relation.

Proof. We have $x \in K \iff \langle x, x \rangle \in H$. ⊣

Thus the halting problem, despite being a precisely formulated problem, is unsolvable. We will see other such problems (i.e., other noncomputable relations). Moreover, there are unsolvable problems in other parts of mathematics. In Chapter 5, we will see that the problem of deciding, given a sentence in arithmetic, whether it is true or false, is unsolvable.

Digression: "Hilbert's tenth problem" is the problem of deciding, given a polynomial equation in many variables with integer coefficients, whether or not it has a solution in the integers. (For example, the equation $x^2 = 9y^2 + 18y + 28$ has the solution $x = 10, y = 2$, but the equation $x^2 = 9y^2 + 18y + 10$ has no solution in the integers.) It is now known (through work of Martin Davis, Yuri Matiyasevich, Hilary Putnam, and Julia Robinson) that this problem is unsolvable.

For a very different example, in symbolic logic, the problem of deciding, given a formula in symbolic logic, whether or not it is true under all interpretations of its symbols, is an unsolvable problem. This result is known as Church's theorem.

In contrast to the halting problem, the ternary relation

$$\{\langle x, y, t \rangle \mid [\![x]\!](y) \downarrow \text{ in } \leq t \text{ steps}\}$$

is primitive recursive:

$$[\![x]\!](y) \downarrow \text{ in } \leq t \text{ steps} \iff (\text{snap}(x, y, t))_0 \geq \text{lh } x$$

Call this ternary relation T, so that

$$T(x, y, t) \iff [\![x]\!](y) \downarrow \text{ in } \leq t \text{ steps}$$
$$\iff (\text{snap}(x, y, t))_0 \geq \text{lh } x.$$

And here we can replace y by $\vec{y}$. That is, for each n, we define the $(n+2)$-ary $T^{(n)}$ as

$$T^{(n)}(x, \vec{y}, t) \iff [\![x]\!]^{(n)}(\vec{y}) \downarrow \text{ in } \leq t \text{ steps}$$
$$\longmapsto (\text{snap}^{(n)}(x, \vec{y}, t))_0 \geq \text{lh } x.$$

and this relation is primitive recursive. (The relation is closely related to one that is often called "the Kleene T-predicate.")

4.1 Recursively Enumerable Relations

The set K, while noncomputable, is not all bad. Although we lack a decision procedure for it, we do have an "acceptance" procedure. That is, its semicharacteristic function

$$c_K(x) = \begin{cases} 1 & \text{if } x \in K \\ \uparrow & \text{if } x \notin K \end{cases}$$

is a computable partial function because

$$c_K(x) = 0 \cdot \mu t\, T(x, x, t) + 1.$$

(Here we are exploiting the fact that the product $0 \cdot \mu t\, T(x, x, t)$ is undefined unless both factors are defined.) So despite the fact that K is not a decidable set, it is at least semidecidable.

We will be interested in other such sets. The following theorem lists four ways of characterizing them.

Theorem: *For an m-ary relation R on* $\mathbb{N}$, *the following conditions are equivalent:*

(a) *The semicharacteristic function of R*

$$c_R(\vec{x}) = \begin{cases} 1 & \text{if } \vec{x} \in R \\ \uparrow & \text{if } \vec{x} \notin R \end{cases}$$

is a computable partial function. (Informally, this condition tells us that we have an effective "acceptance procedure" for R, so that R is an effectively recognizable relation.)

(b) *R is the domain of some computable partial function.*

(c) *For some* $(m + 1)$-*ary computable relation Q,*

$$\vec{x} \in R \iff \exists y\, Q(\vec{x}, y).$$

(We say that R is a Σ_1 *relation if this condition holds. We can think of y as providing "evidence" that* $\vec{x}$ *belongs to R. Geometrically, we can view R as the projection of the relation Q from* $\mathbb{N}^{m+1}$ *to* $\mathbb{N}^m$.)

(d) *For some k and some* $(m + k)$-*ary computable relation Q,*

$$\vec{x} \in R \iff \exists y_1 \cdots \exists y_k\, Q(\vec{x}, y_1, \dots, y_k).$$

Proof. To show equivalence of the conditions, it suffices to obtain four implications, forming a loop. But instead, we will obtain six.

(a) $\Rightarrow$ (b): Easy; $R = \text{dom}\, c_R$.

(b) $\Rightarrow$ (a): $c_{\text{dom} f}(\vec{x}) = 0 \cdot f(\vec{x}) + 1$. That is, whenever f is a computable partial function, then the function mapping $\vec{x}$ to $0 \cdot f(\vec{x}) + 1$ is a computable partial function with the same domain and with range at most $\{1\}$. (By the rules for composition of partial functions, a product such as $0 \cdot f(\vec{x})$ is defined only if both factors are defined.)

(b) $\Rightarrow$ (c): Assume that R is the domain of the computable partial function $[\![e]\!]^{(n)}$. Apply the normal form theorem:

$$\vec{x} \in \text{dom}\, [\![e]\!]^{(n)} \iff \exists t\, [[\![e]\!]^{(n)}(\vec{x}) \downarrow \text{ in } \leq t \text{ steps}]$$
$$\iff \exists t\, T^{(n)}(e, \vec{x}, t)$$

This shows a bit more than (c) states: It shows that in (c), we can get Q to be not only computable but even primitive recursive. And later on, we will want to make use of this extra bit of information. (Here the "evidence" that $\vec{x}$ belongs to R is the *time* at which we discover the fact.)

(c) $\Rightarrow$ (b): Assume that $R(\vec{x}) \Leftrightarrow \exists y\, Q(\vec{x}, y)$ and define

$$f(\vec{x}) = \mu y\, Q(\vec{x}, y).$$

Then f is a computable partial function, and its domain is R.

(c) $\Rightarrow$ (d): Obvious.

(d) $\Rightarrow$ (c): We use the following technique to "collapse quantifiers":

$$\exists y_1 \cdots \exists y_k\, Q(\vec{x}, y_1, \ldots, y_k) \iff \exists y\, Q(\vec{x}, (y)_1, \ldots, (y)_k).$$

The $(m+1)$-ary relation

$$\{\langle \vec{x}, y \rangle \mid Q(\vec{x}, (y)_1, \ldots, (y)_k)\}$$

is computable by the substitution property from page 48. $\dashv$

If R meets the conditions listed in this theorem, we say that R is *recursively enumerable*, abbreviated r.e., or that R is *computably enumerable*, abbreviated c.e. Both the "r.e." and the "c.e." terminologies are in common use. When said aloud, the phrase "r.e." is more euphonious than the phrase "c.e." is. Church's thesis tells us the concept of being a recursively enumerable relation corresponds to the informal idea of being a semidecidable relation. Whenever x belongs to an r.e. set W_e, then we can effectively verify this fact by running program number e on input x until it halts, as it eventually must. (But if $x \notin W_e$, this procedure will run forever, leaving us waiting in vain for an answer, never sure whether to give up or to wait just a bit more.)

For example, any computable relation (and by now, we know many of these) is also recursively enumerable. (It might be a good idea to stop and check that for a computable relation R, each of the conditions (a)–(d) of the preceding theorem holds.) Beyond that, the set K is a recursively enumerable set, and the halting relation H is a recursively enumerable binary relation. (Right?)

The enumeration theorem gives us an enumeration of the r.e. relations. That is, define

$$W_e^{(n)} = \operatorname{dom} \|e\|^{(n)}.$$

Then

$$W_0^{(n)}, \quad W_1^{(n)}, \quad W_2^{(n)}, \quad \ldots$$

is a complete list (with repetitions) of all the recursively enumerable n-ary relations. As usual, when $n = 1$, we can omit the superscript. Thus,

$$W_0, \ W_1, \ W_2, \ \ldots$$

is a complete list of all the recursively enumerable sets of natural numbers.

For a nonexample, consider the set

$$\overline{K} = \{x \mid [\![x]\!](x) \uparrow\}.$$

This is the complement of an r.e. set (such sets are sometimes called co-r.e. sets), but the set $\overline{K}$ is *not* recursively enumerable.

Proposition: *For any recursively enumerable subset W_e of $\overline{K}$, we have $e \in \overline{K} \setminus W_e$ (i.e., $e \in \overline{K}$ but $e \notin W_e$).*

That is, whenever $W_e \subseteq \overline{K}$, then the number e itself is a witness to the fact that equality does not hold. So the proposition immediately implies that $\overline{K}$ is not r.e.

Proof. We have $W_e \subseteq \overline{K}$. It is not possible to have $e \in K$ because that would imply that $e \in W_e \subseteq \overline{K}$. Hence, we can be sure that $e \in \overline{K}$. And from that we get $e \notin W_e$. ⊣

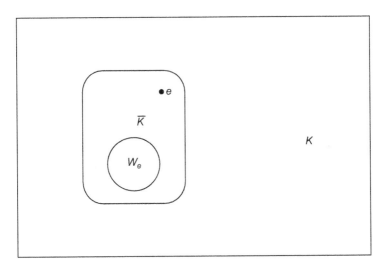

The fact that $\overline{K}$ is not r.e. also follows from the following result.

Kleene's theorem: *A relation is computable if and only if both it and its complement are recursively enumerable.*

We have encountered this theorem before; see page 9. The result appeared in a 1943 article by Kleene; it was observed independently by Post and by Andrzej Mostowski.

Proof. In the one direction, assume that R is a computable relation. Then we know that its complement (with respect to $\mathbb{N}^n$) $\overline{R}$ is also a computable relation. And because all computable relations are also r.e., it follows that both R and $\overline{R}$ are recursively enumerable.

For the more serious direction, we assume that R is an n-ary relation such that both R and $\overline{R}$ are recursively enumerable. Thus each is Σ_1: For some $(n+1)$-ary computable relations P and Q, we have both

$$\vec{x} \in R \iff \exists y\, P(\vec{x}, y) \quad \text{and}$$
$$\vec{x} \in \overline{R} \iff \exists y\, Q(\vec{x}, y).$$

Let

$$f(\vec{x}) = \mu y[\text{either } P(\vec{x}, y) \text{ or } Q(\vec{x}, y)].$$

Thus f searches for evidence, one way or the other. Then f is computable and total (because there always is such a number y). And

$$\vec{x} \in R \iff P(\vec{x}, f(\vec{x}))$$

which tells us that R is a computable relation (by using substitution).   ⊣

Example: We know that the halting relation H

$$\langle x, y \rangle \in H \iff [\![x]\!](y) \downarrow$$

is *not* computable. But H is a recursively enumerable relation because

$$[\![x]\!](y) \downarrow \iff \exists t\, T(x, y, t).$$

Applying Kleene's theorem, we can conclude that the nonhalting relation $\overline{H}$

$$\langle x, y \rangle \in \overline{H} \iff [\![x]\!](y) \uparrow$$

is *not* recursively enumerable.

It is not hard to see that the union of two n-ary recursively enumerable relations is again recursively enumerable. And the same is true for intersections. (See Exercise 4.) But the complement of an r.e. relation is not r.e., unless the relation is computable.

For an n-place partial function f, its *graph* is the $(n + 1)$-ary relation

$$\{\langle \vec{x}, y \rangle \mid f(\vec{x}) = y\}.$$

In fact, a standard procedure is to define a function to be a set of ordered pairs with a certain single-valuedness property. In this approach, a function simply *is* its graph. We will ignore this point.

Theorem: *A partial function is a computable partial function if and only if its graph is a recursively enumerable relation.*

We observed earlier (page 48) that the graph of a *total* computable function is a computable relation. This can fail in the nontotal case; for example, the graph of c_K, the semicharacteristic function of K, is a noncomputable binary relation because

$$x \in K \iff \langle x, 1 \rangle \in \text{ the graph of } c_K.$$

Proof. In one direction, assume that f is a partial function whose graph is the Σ_1 relation

$$\{\langle \vec{x}, y \rangle \mid \exists z\, R(\vec{x}, y, z)\},$$

where R is a computable relation. Then given $\vec{x}$, we need a "two-dimensional" search: we want to locate both the answer y and the evidence z. The μ-operator does the search; the dimensionality is easy to deal with:

$$f(\vec{x}) = (\mu t\, R(\vec{x}, (t)_0, (t)_1))_0$$

That is, we search for y and z, and then we ignore z and return y. This equation shows that f is a computable partial function.

In the other direction, consider the computable partial function $[\![e]\!]^{(n)}$. We apply the normal form theorem:

$$
\begin{aligned}
\langle \vec{x}, y \rangle \in \text{ the graph of } [\![e]\!]^{(n)} &\iff \exists t[[\![e]\!]^{(n)}(\vec{x}) = y \text{ in } \leq t \text{ steps}] \\
&\iff \exists t[T^{(n)}(e, \vec{x}, t) \quad \text{and} \\
&\qquad ((\text{snap}(e, \vec{x}, t)_1))_0^* = y]
\end{aligned}
$$

Observe that this relation is Σ_1.   ⊣

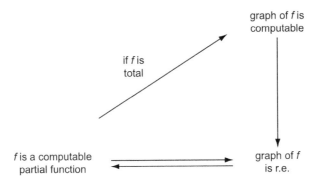

In the foregoing proof, we have made use of the fact that

$$T^{(n)}(e, \vec{x}, t) \quad \text{and} \quad ((\text{snap}(e, \vec{x}, t)_1))_0^* = y$$

is a primitive recursive condition on $e, \vec{x}, t, y$ that says "$[\![e]\!]^{(n)}(\vec{x}) = y$ in $\leq t$ steps." The fact that this condition is primitive recursive will be useful later as well.

We know that the domain of a computable partial function is r.e. The same is true of the range.

Theorem: *The range of any computable partial function is a recursively enumerable subset of* $\mathbb{N}$.

Proof 1. The range of a function f is the set

$$\{y \mid \exists \vec{x} \, \langle \vec{x}, y \rangle \in \text{ the graph of } f\}.$$

By the preceding theorem, we can express the graph of f as a Σ_1 relation

$$\{\langle \vec{x}, y \rangle \mid \exists t \, Q(\vec{x}, y, t)\},$$

where Q is a computable relation. Then

$$y \in \text{ran} f \iff \exists \vec{x} \, \exists t \, Q(\vec{x}, y, t),$$

which shows that the range is r.e.   ⊣

Proof 2. The same argument can be condensed into one line:

$$y \in \text{ran} f \iff \exists \vec{x} \, \exists t [f(\vec{x}) = y \text{ in } \leq t \text{ steps}].$$

⊣

What is especially "enumerable" about recursively enumerable sets? The following theorem (which also provides a converse to the preceding theorem) gives an answer.

Theorem: *A subset of $\mathbb{N}$ is recursively enumerable if and only if it either is empty or is the range of a total computable one-place function.*

Proof. In one direction, the preceding theorem applies: The range of a total computable one-place function is r.e., as is the range of *any* computable partial function. And the empty set is both computable and recursively enumerable.

It is the other direction that requires proof. Assume that A is a nonempty computable enumerable subset of $\mathbb{N}$; say c is some particular member of A. Then, A is Σ_1; say

$$x \in A \iff \exists y \, Q(x, y)$$

for computable Q. Then a two-place function with range equal to A is the function f_2:

$$f_2(x, y) = \begin{cases} x & \text{if } Q(x, y) \\ c & \text{otherwise.} \end{cases}$$

But we want a one-place function. Define

$$f(t) = \begin{cases} (t)_0 & \text{if } Q((t)_0, (t)_1) \\ c & \text{otherwise.} \end{cases}$$

(Informally, if t says to us, "I have found a member of A and here it is and here is the evidence," then we act accordingly.) Then f is a total computable function (by definition-by-cases) and its range is A. In fact, because we can get Q to be primitive recursive, we can get f to be a primitive recursive function. (In particular, this shows that the range of a primitive recursive function need *not* be a primitive recursive set, or even a computable set.)   ⊣

In this theorem (in the more interesting direction), we have

$$A = \{f(0), f(1), f(2), \ldots\}.$$

That is, the function f gives us an effective *enumeration* of the set A. (Julia Robinson suggested using the word "listable" instead of "enumerable.") It is not, in general, possible to get f to enumerate the members of A in increasing order, unless A is computable. (See the exercises.)

The preceding theorems show that

$$\text{ran } [\![0]\!], \quad \text{ran } [\![1]\!], \quad \text{ran } [\![2]\!], \quad \ldots$$

give a complete list (with repetitions) of exactly the recursively enumerable subsets of $\mathbb{N}$. The advantage of using the domain instead of the range (as we did in defining W_e) is that it extends to recursively enumerable relations that are subsets of $\mathbb{N}^k$ for larger k.

Among the computable partial functions, the ones that are *total* have a particular "user-friendly" status: When you give an input to a total function, you get back an output. Define the set

$$\text{Tot} = \{e \mid [\![e]\!] \text{ is total}\}$$

of indices of total one-place computable functions. This is not a computable set, as shown in the exercises. And moreover, it is not even a recursively enumerable set. The following theorem proves an even stronger fact.

Theorem: *Assume that A is a recursively enumerable subset of Tot. Then, there is some total one-place computable function that does not equal $[\![a]\!]$ for any number a belonging to A.*

Proof. We can obviously assume that A is nonempty. Hence, by the preceding theorem, A is the range of some total computable one-place function g. Define the following function:

$$f(x) = [\![g(x)]\!](x) + 1.$$

(Alternatively, we can just as well use $f(x) = 1 \,\dot-\, [\![g(x)]\!](x)$.) This is a computable partial function. Moreover, it is total, because $g(x) \in \text{Tot}$ for all x. Could f equal $[\![a]\!]$ for some a in A? We know that $a = g(t)$ for some t. But then,

$$[\![a]\!](t) = [\![g(t)]\!](t) \text{ and } f(t) = [\![g(t)]\!](t) + 1$$

and these are defined. So f cannot be the same as $[\![a]\!]$; the two functions differ at t. ⊣

To illustrate this theorem, suppose that you are teaching a beginning programming course. Your students keep turning in programs for nontotal functions, which you find annoying. So you give them a set of rules such that the rule-following programs always compute total functions. Assume that the set of rule-following programs is computable, or at least recursively enumerable (so we can recognize a rule-following program, when we see one). Then it follows from the above theorem that some total function will have *no* rule-following program. That is, imposing the rules has had the unintended side effect of limiting the class of total programmable functions. (Programming rules may, indeed, be a good thing. But one should be aware that there is a cost – there may be some total functions that can be programmed *only* by breaking the rules.)

For a second illustration, suppose we have a program that we are pretty sure computes a total function. But to be certain, we want a *correctness proof*. Will there necessarily be a proof that the program halts on all possible inputs?

This question leads us to ask: What is a proof? Because a proof should be a finite string of symbols, we can code a proof by a number, in much the same way as we coded programs by numbers. Suppose that we have formalized a notion of "proof of totality" in such a way that the following two assumptions are met.

1. Proofs don't lie. That is, whenever there exists a proof that program number y is total, then $[\![y]\!]$ is really total.
2. We can effectively recognize proofs. That is, the relation

$$\{(y, z) \mid z \text{ codes a proof that program } y \text{ is total}\}$$

   is recursively enumerable. (The idea behind this assumption is that if a proof is to *convince* someone, then that person must at the very least be able to *verify* the proof in an effective way.)

Then apply the theorem to the set

$$B = \{y \mid \exists z(z \text{ codes a proof that program } y \text{ is total})\}$$

of provably total programs (in this proof system). By the first assumption, $B \subseteq$ Tot. By the second assumption, B is recursively enumerable. We conclude that there exist total computable functions such that *no* programs for those functions are provably total in this system!

(The fact that Tot is not r.e. implies that some total programs are not provably total in the system. The result here is stronger. It states that there are computable total functions *all* of whose programs fail to be provably total in the system.)

There is a connection here to Gödel's incompleteness theorem, which we will encounter in Chapter 5.

Next we want to show that the class of recursively enumerable relations is closed under substitution of computable functions.

Proposition:

(a) *If A is a recursively enumerable subset of $\mathbb{N}$ and f is a total computable function, then $\{x \mid f(x) \in A\}$ is also recursively enumerable.*

(b) *If R is a recursively enumerable n-ary relation and $f_1, \ldots, f_n$ are total computable m-place functions, then the m-ary relation*

$$\{\vec{x} \mid R(f_1(\vec{x}), \ldots, f_n(\vec{x}))\}$$

is also recursively enumerable.

In fact, the functions here do not even need to be total, if, for example, $\{x \mid f(x) \in A\}$ is understood as the set of x's for which $f(x)$ is defined and belongs to A. So let's restate the proposition of greater generality. We write "$f(x) \downarrow$" to indicate that $f(x)$ is defined (i.e., $x \in \text{dom} f$).

Proposition:
(a) *If A is a recursively enumerable subset of $\mathbb{N}$ and f is a computable partial function, then $\{x \mid f(x) \downarrow$ and $f(x) \in A\}$ is also recursively enumerable.*
(b) *If R is a recursively enumerable n-ary relation and $f_1, \ldots, f_n$ are computable partial m-place functions, then the m-ary relation*

$$\{\vec{x} \mid \langle f_1(\vec{x}), \ldots, f_n(\vec{x})\rangle \downarrow \text{ and belongs to } R\}$$

is also recursively enumerable.

Proof. (b) We know that the semicharacteristic function c_R of R is a computable partial function. The semicharacteristic function of the new relation is

$$\vec{x} \mapsto c_R(f_1(\vec{x}), \ldots, f_n(\vec{x})),$$

which is a composition of computable partial functions. $\dashv$

But note that for a *computable* set A and a computable partial function f, the set $\{x \mid f(x) \downarrow$ and $f(x) \in A\}$ is not computable, in general. For example, if $A = \mathbb{N}$, then this set is simply the domain of f, which might not be computable.

For a computable relation Q and a computable partial function g, we have seen that the partial function g^Q defined by cases

$$g^Q(\vec{x}) = \begin{cases} g(\vec{x}) & \text{if } \vec{x} \in Q \\ 0 & \text{if } \vec{x} \notin Q \end{cases}$$

is again a computable partial function. If we weaken the assumption on Q to assume only that Q is recursively enumerable, then g^Q might fail to be a computable partial function (for example, if $Q = K$ and $g(x) = 1$). But we do have the following result.

Proposition (Definition by semicases): *Assume that g is a computable partial n-place function and Q is a recursively enumerable n-ary relation. Then the function*

$$(g \upharpoonright Q)(\vec{x}) = \begin{cases} g(\vec{x}) & \text{if } \vec{x} \in Q \\ \uparrow & \text{if } \vec{x} \notin Q \end{cases}$$

is a computable partial function.

Note that $(g \restriction Q)(\vec{x})$ is undefined unless both $\vec{x} \in Q$ and $g(\vec{x}) \downarrow$. Informally, the procedure for computing $(g \restriction Q)(\vec{x})$ involves first trying to verify that $\vec{x} \in Q$, and then computing $g(\vec{x})$.

Proof 1. $(g \restriction Q)(\vec{x}) = g(\vec{x}) + 0 \cdot c_Q(\vec{x})$. $\dashv$

Proof 2. $(g \restriction Q)(\vec{x}) = y \Leftrightarrow \vec{x} \in Q$ and $g(\vec{x}) = y$, so the graph of $g \restriction Q$ is the intersection of two $(n+1)$-ary r.e. relations. $\dashv$

All of the noncomputable sets are noncomputable, but some are more noncomputable than others. One way to make sense out of that statement is to look at ways in which membership questions about one set might be "reduced" to membership questions about another.

More specifically, for sets A and B of natural numbers, we say that A is *many-one reducible* to B (in symbols, $A \leq_m B$) if there exists some total computable function such that

$$x \in A \iff f(x) \in B$$

for all x. That is, the function f is, in a sense, effectively reducing the question whether $x \in A$ to a question about B.

Note that if $A \leq_m B$, then it is automatically true that $\overline{A} \leq_m \overline{B}$, by using the same function.

Digression: The name "many-one reducible" in not particularly informative. The name "mapping reducible" has been suggested as an alternative. And it retains the feature of starting with the letter "m" so that the symbol $\leq_m$ does not need to be altered.

Proposition: *Assume that A and B are sets of natural numbers with $A \leq_m B$.*

(a) *If B is a computable set, then A is also computable.*
(b) *If B is recursively enumerable, then A is also recursively enumerable.*

Proof. Part (a) is a substitution rule, which we already have seen back on page 48: $A = \{x \mid f(x) \in B\}$.

Part (b) follows from an earlier proposition. If f is a computable function that many-one reduces A to B, then

$$A = \{x \mid f(x) \in B\},$$

and so if B is recursively enumerable, then so is A. $\dashv$

For example, in order to show that some set is not recursively enumerable, it suffices to show that $\overline{K}$ is many-one reducible to it.

Exercises

1. Obviously Tot $\subseteq K$. Show that there is no computable set A with Tot $\subseteq A \subseteq K$. Suggestion: Consider the function defined by the equation:

$$f(x) = \begin{cases} [\![x]\!](x) + 1 & \text{if } x \in A \\ 0 & \text{if } x \notin A. \end{cases}$$

2. (a) Assume that f is a total computable one-place nondecreasing function (that is, $f(x) \leq f(x + 1)$ for all x). Further assume that the range of f is infinite. Show that the range of f is a computable set.

   (b) Suppose that in part (a), we drop the assumption that the range is infinite. Show that the same conclusion still holds.

3. Assume that A is an infinite recursively enumerable set of natural numbers. Show that there is a total computable strictly increasing function g (that is, $g(x) < g(x + 1)$ for all x) whose range is included in A. (It follows from this exercise that every infinite recursively enumerable set has an infinite computable subset.)

4. (a) Show that the intersection of two n-ary recursively enumerable relations is again recursively enumerable.

   (b) Show that the union of two n-ary recursively enumerable relations is again recursively enumerable.

5. Assume that f is a total computable function. Show that

$$\bigcup_{n \in \mathbb{N}} W_{f(n)}$$

   is recursively enumerable. (That is, a computably indexed union of r.e. sets is r.e.)

6. (a) Show that there is a computable partial function f such that whenever $W_x \neq \emptyset$, then $f(x) \downarrow$ and $f(x) \in W_x$. (That is, the function f finds *some* member of W_x, provided there is a member.)

   (b) Assume that R is a recursively enumerable binary relation. Construct a computable partial function f such that whenever $\exists y \, R(x, y)$, then $f(x) \downarrow$ and $R(x, f(x))$. (That is, the function f finds *some* y such that $R(x, y)$, provided there is one.)

7. Show that the set $\{x \mid [\![x]\!](x) = 0\}$ (i.e., the set of x's for which $[\![x]\!](x)$ is defined and equals 0) is recursively enumerable but not computable.

8. Let

$$A = \{x \mid [\![x]\!](x) = 0\} \quad \text{and} \quad B = \{x \mid [\![x]\!](x) = 1\}.$$

   (a) Show that A and B are disjoint r.e. sets.

   (b) Show that A and B are *computably inseparable*, that is, there is no computable set D with $A \subseteq D$ and $B \subseteq \overline{D}$. Suggestion: Suppose we had such a set D; let d be an index of its characteristic function. What can you say about $[\![d]\!](d)$?

9. Give an alternative proof of Kleene's theorem as follows. Assume that R is a relation for which both R and $\overline{R}$ are recursively enumerable. Show that C_R, the characteristic function of R, has a recursively enumerable graph.

4.2 Parameters

Next we want to turn our attention to the subject of *calculating programs.* That is, suppose we want a program that will meet some particular need. Possibly we have a good reason to know that a program *exists* that meets the need. But we might want more than that; we might want actually to *find* such a program.

For example, we know that every total constant function is computable. (As noted in item **2** in Chapter 2, every such function is primitive recursive.) But even more is true. Given a constant k, we can actually compute an index of the one-place function that is constantly equal to k. See Exercise 4 in Chapter 3.

For another example, we have recently seen that the range of any computable partial one-place function $[\![e]\!]$ is an r.e. set, and therefore is W_y for some y. That is, there exists some index y of a function whose domain is the set we want. But a stronger fact is true: Given e, we can actually compute such a number y. (We will see a proof of this later.) That is, there is a computable function f such that

$$\operatorname{ran}[\![e]\!] = W_{f(e)}$$

for every e.

For a third example, suppose we have a two-place computable partial function f. Then, clearly the one-place function g obtained by holding the second variable fixed as a parameter, say,

$$g(x) = f(x, 3),$$

is a computable partial function; we have merely applied composition with a computable (and constant) function. Or in terms of register machines, we can make a program for g that increments register 2 three times, and then follows the program for f.

But by standing back and looking at the previous paragraph, we perceive a more subtle fact. We were able to find explicitly a program for g, given the parameter 3 and a program for f. So there should be a computable function ρ that, given an index e for f and given the parameter 3, will compute an index of the function g:

$$f(x, 3) = [\![\rho(e, 3)]\!](x)$$

Not only does there *exist* a program to compute g, but, given the parameter 3, we can actually lay our hands on such a program.

Parameter theorem: *There is a primitive recursive function ρ such that the equation*

$$[\![e]\!]^{(2)}(x, y) = [\![\rho(e, y)]\!](x)$$

holds for all e, x, and y. (Here equality has the usual meaning: either both sides are undefined, or both sides are defined and are the same.) Moreover, ρ is one-to-one.

The idea is that if we have a computable partial two-place function $[\![e]\!]^{(2)}$, and we want to hold the last variable fixed as a parameter, then ρ will actually calculate an index for the resulting one-place function.

Proof, first try. The program that increments register 2 exactly y times has the Gödel number

[#I2, #I2, . . . , #I2].

The Gödel number of the instruction I2 is $[0, 2] = 2^1 \cdot 3^3 = 54$. So the Gödel number of the program incrementing register 2 exactly y times is

$$[54, 54, \dots, 54] = \prod_{t < y} p_t^{55},$$

which is a primitive recursive function of y; call it $k_2(y)$. This suggests that we might be able to take

$$\rho(e, y) = k_2(y) * e$$

using the concatenation function $*$ from item **21** in Chapter 2. This will indeed work for "nice" values of e. But difficulties can arise if e is the Gödel number of a program that makes bad negative jumps (i.e., jumps to a point before the start of the program). Also, when e is not the Gödel number of a program at all, then we are less certain that this equation for ρ will give us the theorem that we are after.

Proof, second try. There is a way to circumvent the difficulties that arise in the first try. We use a universal function instead. The universal partial function

$$\Phi^{(2)}(e, x, y) = [\![e]\!]^{(2)}(x, y)$$

is a computable partial function; fix some register-machine program Q that computes it. (By the way we defined the verb "computes," the program Q always either runs forever or comes to a good halt. Thus, we can avoid the difficulties mentioned earlier.) Define

$$\rho(e, y) = k_2(y) * k_3(e) * q,$$

where q is the Gödel number of the program Q. Here k_3 is the function like k_2, but it uses register 3. Clearly ρ is primitive recursive. (Here q is a fixed constant.)

To check that ρ works, let's calculate $[\![\rho(e, y)]\!](x)$. Here $\rho(e, y)$ is the Gödel number of a program, and we know what that program does, given the input x:

- First it inserts y into register 2.
- Secondly, it inserts e into register 3.
- Finally, it runs Q to try to find $\Phi^{(2)}(e, x, y) = [\![e]\!]^{(2)}(x, y)$, if this is defined.

That is, we get exactly $[\![e]\!]^{(2)}(x, y)$ if this is defined, and we get nothing if this is undefined. So the equation

$$[\![e]\!]^{(2)}(x, y) = [\![\rho(e, y)]\!](x)$$

holds.

The function ρ is one-to-one because different values of e and y will result in different programs, and hence different Gödel numbers.   ⊣

As an example of an application of the parameter theorem, we can "uniformize" an earlier result. We have seen that the range of any computable partial function $[\![r]\!]$ is an r.e. set. The "uniformized" version of this statement is that there is some total computable (in fact, primitive recursive) function f such that

$$\operatorname{ran}[\![r]\!] = W_{f(r)}$$

for every r. That is, the function f goes out and finds an "r.e. index" for the range of $[\![r]\!]$.

We know that

$$y \in \operatorname{ran}[\![r]\!] \iff \exists x [\![r]\!](x) = y$$
$$\iff \exists x \exists t [\![r]\!](x) = y \text{ in } t \text{ steps.}$$

Look at the function that searches for x and t here:

$$g(y, r) = \mu s[[\![r]\!]((s)_0) = y \text{ in } (s)_1 \text{ steps}].$$

Then g is a computable partial function, so it is $[\![e]\!]^{(2)}$ for some e. Parameterize out r:

$$[\![\rho(e, r)]\!](y) = [\![e]\!]^{(2)}(y, r) = g(y, r) = \mu s[[\![r]\!]((s)_0) = y \text{ in } (s)_1 \text{ steps}].$$

This quantity is defined if and only if there is some s to be found, which happens if and only if $y \in \operatorname{ran}[\![r]\!]$. That is,

$$\operatorname{ran}[\![r]\!] = W_{\rho(e,r)},$$

which is what we wanted.

Here is another application of the parameter theorem:

Lemma: *Assume that S is an r.e. set of natural numbers, and that f is a computable partial one-place function. Then there is a one-to-one primitive recursive function g such that for any y:*

(i) *If $y \in S$, then $[\![g(y)]\!]$ is the partial function f.*
(ii) *If $y \notin S$, then $[\![g(y)]\!]$ is the empty function, that is, the function that is undefined for all inputs.*

So the function g produces indices of functions. In fact, $g(y)$ is an index either for f or for the empty function. And which of these two alternatives occurs is determined by whether or not $y \in S$.

The proof involves looking at the two-place partial function

$$h(x, y) = \begin{cases} f(x) & \text{if } y \in S \\ \uparrow & \text{if } y \notin S, \end{cases}$$

which can be computed by the one-line instruction,

"First verify that $y \in S$ and secondly find $f(x)$."

The point is that given y, we can effectively find the above one-line instruction. There is a primitive recursive function that, given y, produces the above line with the value y filled in. The actual proof cleans this argument up.

Proof. Applying definition by semicases, we obtain a two-place computable partial function h:

$$h(x, y) = \begin{cases} f(x) & \text{if } y \in S \\ \uparrow & \text{if } y \notin S. \end{cases}$$

So h is $[\![e]\!]^{(2)}$ for some number e. Now parameterize out the y. We get

$$[\![\rho(e, y)]\!](x) = [\![e]\!]^{(2)}(x, y) = \begin{cases} f(x) & \text{if } y \in S \\ \uparrow & \text{if } y \notin S. \end{cases}$$

Now let $g(y) = \rho(e, y)$ for this number e. Then g is primitive recursive, one-to-one, and

$$[\![g(y)]\!](x) = [\![\rho(e, y)]\!](x) = \begin{cases} f(x) & \text{if } y \in S \\ \uparrow & \text{if } y \notin S. \end{cases}$$

Thus, the partial function $[\![g(y)]\!]$ either is f (if $y \in S$) or else is the empty function (if $y \notin S$).   ⊣

Application: We can show that $K \leq_m \text{Tot}$. In the lemma, take S to be K and take f to be the identity function I_1^1 (or any total computable function). We obtain a primitive recursive function g such that whenever $y \in K$, then $g(y) \in \text{Tot}$ because $g(y)$ is an index of the total function I_1^1. And whenever $y \notin K$, then $g(y) \notin \text{Tot}$ because $g(y)$ is an index of the empty function, which is certainly not total.

It follows that $K \leq_m \text{Tot}$ under this function g. And so automatically $\overline{K} \leq_m \overline{\text{Tot}}$ under the same function. Consequently, $\overline{\text{Tot}}$ is not recursively enumerable. (We saw earlier that Tot itself is not recursively enumerable.)

Similarly, the set of indices of the empty function

$$\{e \mid [\![e]\!] \text{ is empty}\} = \{e \mid W_e = \emptyset\}$$

is not recursively enumerable because $\overline{K}$ is many-one reducible to it by the same function g.

Application: We can show that $S \leq_m K$ for any r.e. set S. Apply the lemma where f again is I_1^1 or some other total computable function. We obtain a one-to-one primitive recursive function g such that for any y,

$$y \in S \Rightarrow [\![g(y)]\!] \text{ total} \Rightarrow g(y) \in K$$
$$y \notin S \Rightarrow [\![g(y)]\!] \text{ empty} \Rightarrow g(y) \notin K.$$

Thus, $S \leq_m K$ under g.

Definition: A *complete* recursively enumerable set is a recursively enumerable subset C of $\mathbb{N}$ with the property that

$$A \leq_m C$$

for every recursively enumerable subset A of $\mathbb{N}$.

The preceding application proves the following result.

Proposition: *The set K is a complete recursively enumerable set.*

Here is another complete r.e. set obtained in a more direct way:

$$C = \{x \mid (x)_0 \in W_{(x)_1}\}$$

Then, C is recursively enumerable because

$$x \in C \iff \exists t \, T((x)_1, (x)_0, t).$$

And C is complete because

$$x \in W_a \iff [x, a] \in C.$$

We can "vectorize" the parameter theorem as follows. (In this form, the theorem is commonly called the "*S-m-n* theorem," for no very good reason.)

Parameter theorem: *For each m and n, there is an $(n + 1)$-place primitive recursive function ρ_{mn} such that the equation*

$$[\![e]\!]^{(m+n)}(\vec{x}, \vec{y}) = [\![\rho_{mn}(e, \vec{y})]\!]^{(m)}(\vec{x})$$

for all e, all m-tuples $\vec{x}$, and all n-tuples $\vec{y}$. (Here equality has the usual meaning: either both sides are undefined, or both sides are defined and are the same.) Moreover, ρ_{mn} is one-to-one.

Proof. We proceed as before. Here $\vec{x}$ is $\langle x_1, \ldots, x_m \rangle$ and $\vec{y}$ is $\langle y_1, \ldots, y_n \rangle$.

$$\rho_{mn}(e, y_1, \ldots, y_n) = k_{m+1}(y_1) * \cdots * k_{m+n}(y_n) * k_{m+n+1}(e) * q,$$

where q is the Gödel number of our program that computes $\Phi^{(m+n)}$. (Here m, n, and q are fixed; e, $\vec{x}$, and $\vec{y}$ are the variables.)   ⊣

For example, we have long known that the composition $f \circ g$ of two computable partial functions is a computable partial function. But now, we can obtain a "uniform" version of that statement: There is a total computable function h such that

$$[\![h(x, y)]\!] = [\![x]\!] \circ [\![y]\!]$$

for all x and y. That is, not only does a program for the composition exist, but also we can effectively find it.

The property we need h to satisfy is

$$[\![h(x, y)]\!](t) = [\![x]\!]([\![y]\!](t))$$

for all t. So look at the right-hand side as a function of all the variables: it is a computable partial function of t, x, and y. So there is some e for which

$$[\![e]\!]^{(3)}(t, x, y) = [\![x]\!]([\![y]\!](t))$$

for all t, x, and y. We proceed to parameterize out the x and y:

$$[\![e]\!]^{(3)}(t, x, y) = [\![\rho_{12}(e, x, y)]\!](t).$$

So we can take $h(x, y) = \rho_{12}(e, x, y)$ for this e. Then h is a primitive recursive function.

The following result was published by H. Gordon Rice in 1953; it is due independently to Vladimir Uspensky.

Rice's theorem: *Let C be a set of one-place computable partial functions and let $I_C = \{e \mid [\![e]\!] \in C\}$ be the set of all indices of members of C. Then I_C is noncomputable except in the two trivial cases: $C = \emptyset$ (in which case, $I_C = \emptyset$) or C is the set of all one-place computable partial functions (in which case, $I_C = \mathbb{N}$).*

Proof. First we look to see where the empty function (the function that is undefined everywhere), call it $\emptyset$, is.

Case I: The empty function $\emptyset$ is *not* in C. We are given that C has some function in it; say $\psi \in C$. We can apply a recent lemma to obtain a computable total function g with the following properties:

(i) Whenever $y \in K$, then $[\![g(y)]\!]$ is the function ψ.
(ii) Whenever $y \notin K$, then $[\![g(y)]\!]$ is the empty function $\emptyset$.

Then $K \leq_m I_C$ under the function g:

$$y \in K \iff g(y) \in I_C.$$

Hence I_C is not computable.

Case II: The empty function $\emptyset$ *is* in C. We proceed much as before. We are given that not everything is in C; say ψ is a computable partial function not in C. As before, we can get a computable total function g with the following properties:

(i) Whenever $y \in K$, then $[\![g(y)]\!]$ is the function ψ.
(ii) Whenever $y \in \overline{K}$, then $[\![g(y)]\!]$ is the empty function $\emptyset$.

Then $\overline{K} \leq_m I_C$ under the function g:

$$y \in \overline{K} \iff g(y) \in I_C.$$

Hence, I_C is not recursively enumerable.   ⊣

In more informal terms, we can restate Rice's theorem as follows: Assume we have in mind some *property* of one-place computable partial functions. Further assume that this property is nontrivial, in the sense that at least one computable partial function has the property, but not all do. Then, we can conclude that the problem of determining whether or not a given number is the index of a partial function with the property is undecidable.

For example, suppose we focus attention on a particular computable function, such as the doubling function $x \mapsto 2x$. The doubling function can be computed by a very simple program, but it is also computed by some convoluted programs. Rice's theorem tells us that we cannot always decide whether a given number is an index of the doubling function or not. (And in particular, this shows that the doubling function has infinitely many indices, which is not surprising.)

For another example, Rice's theorem shows that

$$\{x \mid W_x \text{ is the set of primes}\}$$

is not computable. We take C to be the set of computable partial functions f for which $\mathrm{dom}\, f$ is the set of primes.

Our proof of Rice's theorem actually shows a bit more than the theorem states. On the one hand, it shows that when the empty function is in C, then I_C is not r.e. And on the other hand, it shows that if the empty function is not in C, then $\overline{K} \leq_m \overline{I_C}$ and hence $\overline{I_C}$ is not r.e. (that is, I_C is not "co-r.e."; it is not the complement of an r.e. set).

For example, taking C to be the collection of total computable functions, we have $I_C = \mathrm{Tot}$. Because the empty function is not total, we see once again that Tot is not co-r.e.

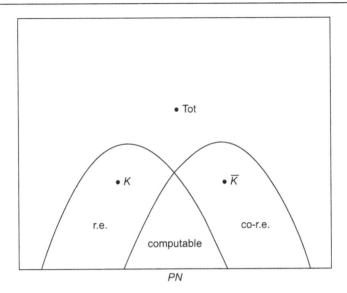

Another consequence of the parameter theorem is the following result, which is due to Kleene.

Recursion theorem: *For any computable partial function g, we can find an e such that*

$$[\![e]\!](x) = g(e, x)$$

for all x.

Again, x can be replaced by an n-tuple $\bar{x}$. The proof of the recursion theorem is very much like the argument generally used in logic to prove Gödel's incompleteness theorem; see Exercise 20. We will not pursue any of these topics right now.

Exercises

10. We know that the product $f \cdot g$ of computable partial functions is a computable partial function. Show that there is a total computable function h such that the equation

$$[\![h(x, y)]\!](t) = [\![x]\!](t) \cdot [\![y]\!](t)$$

holds for all t, x, and y.

11. We know that the union of two recursively enumerable sets is recursively enumerable. Show that there is a total computable function g, such that

$$W_{g(x,y)} = W_x \cup W_y$$

for all x and y.

12. Show that

$$\{t \mid [\![t]\!](0) = 0\} \leq_m \{y \mid [\![y]\!](0) = 7\}.$$

13. **(a)** Show that $K \leq_m \{x \mid W_x$ is infinite$\}$.
   (b) Show that $K \leq_m \{x \mid W_x$ is finite$\}$.
14. Show that Tot $\leq_m \{y \mid W_y$ is infinite$\}$.
15. Show that the binary relation Q defined by the condition

$$\langle x, y \rangle \in Q \iff [\![x]\!] \text{ and } [\![y]\!] \text{ are the same function}$$

is not computable. (This says, roughly, that the problem of determining, given two programs, if they compute the same function, is undecidable.)
16. Let f be some fixed computable partial function. Let I_f be its set of indices:

$$I_f = \{x \mid [\![x]\!] = f\}$$

   (a) Show that I_f is never a computable set. Remark: Suppose you are teaching a programming class, and you assign to your students the problem of writing a program for f. Then the set of correct answers to this problem is an undecidable set!
   (b) Further assume that f is total. Show that Tot $\leq I_f$.
17. Show that the binary relation R defined by the condition

$$\langle x, y \rangle \in R \iff W_x = W_y$$

is not computable.
18. **(a)** Prove a uniformized version of Exercise 2(a). That is, show that there is a primitive recursive function g such that whenever $[\![e]\!]$ is nondecreasing with infinite range, then $g(e)$ is an index for the characteristic function of ran $[\![e]\!]$.
   (b) Show that Exercise 2(b) does not uniformize, even if we add the assumption that the range is finite. That is, show that there cannot be a computable partial function g such that whenever $[\![e]\!]$ is nondecreasing with finite range, then $g(e) \downarrow$ and $g(e)$ is an index for the characteristic function of ran $[\![e]\!]$. Suggestion: Look at the characteristic function of $\{\langle t, x \rangle \mid T(x, x, t)\}$ as a function of t.
19. Show that there is no computable partial function h such that whenever the set W_y is computable, then $h(y) \downarrow$ and $h(y)$ is an index for the characteristic function of W_y. This shows that knowing an acceptance procedure for a set, plus knowing that the set is actually decidable, does not in general lead us to a decision procedure for the set. Suggestion: Look at the function

$$f(u, x) = \begin{cases} \mu t \, T(x, x, t) & \text{if } u = 0 \\ \uparrow & \text{if } u > 0. \end{cases}$$

20. (a) Using the parameter theorem, show that there is a primitive recursive function δ such that the equation

$$[\![\delta(y)]\!](x) = [\![y]\!]^2(x, y)$$

for all x and y. (Here equality means that either both sides are undefined, or both are defined and are the same.)

(b) Now assume that g is a two-place partial computable function. Then the function

$$\langle x, y \rangle \mapsto g(\delta(y), x)$$

is a computable partial function; let q be an index of it, and let $e = \delta(q)$. Show that for this e,

$$[\![e]\!](x) = g(e, x)$$

for all x, thereby proving the recursion theorem.

21. Show that there is a program so narcissistic that it outputs only its own Gödel number. That is, show that for some number e, the equation

$$[\![e]\!](x) = e$$

holds for all x.

22. Let $S = \{x \mid [\![x]\!](3) = 24\}$. Is S a computable set? Is S r.e.? Is $\overline{S}$ r.e.?

23. Show that the set

$$\{x \mid [\![x]\!](t) \uparrow \ \text{for all } t \leq 900\}$$

is not r.e.

24. Show that there is no computable partial function f such that whenever W_x is nonempty, then $f(x)$ is defined and is the *least* member of W_x.

5 Connections to Logic

In this chapter,[1] we want to connect concepts of computability with concepts of *definability*. The idea of definability comes from logic. Roughly speaking, one can specify a language and then study what sets or relations might have exact definitions that can be formulated in that language.

In our case, we will take a language for the arithmetic of natural numbers (that is, number theory). One goal will be to show that every computable set is definable in this language.

This connection between computability and definability has some interesting consequences. For one, it will show that the set of true sentences of arithmetic is very far from being a computable set. And for another, we will come to Gödel's (first) incompleteness theorem. This theorem says that starting from any computable set of true axioms, one cannot possibly hope to derive all the true sentences of arithmetic.

Historically, Gödel's incompleteness theorem (1931) preceded by five years the beginnings of computability theory. But there is an advantage to running history backwards and looking at Gödel's theorem from the point of view of computability theory.

First, however, we want to build on our characterization of recursively enumerable sets as being the Σ_1 sets. The notation "Σ_1" already suggests that there ought be a generalization to Σ_2 and then to Σ_n.

5.1 Arithmetical Hierarchy

We have defined (on page 82) a relation R on the natural numbers to be Σ_1 if for some computable relation Q, we have

$$\vec{t} \in R \iff \exists x \, Q(\vec{t}, x)$$

for all $\vec{t}$. That is, a Σ_1 relation might not be computable, but it is only one quantifier away from computability. We now want to extend this measurement of "distance away from computability."

Define R to be Π_1 if for some computable relation Q, we have

$$\vec{t} \in R \iff \forall x \, Q(\vec{t}, x)$$

for all $\vec{t}$. For example, the set $\overline{K}$ is a Π_1 set because

$$e \in \overline{K} \iff \forall y \, \overline{T}(e, e, y),$$

---

[1] Chapters 5, 6, and 7 are largely independent and can be read in any order.

where T is the ternary relation

$$T(x, v, t) \iff [\![x]\!](v) \downarrow \text{ in } \le t \text{ steps,}$$

which we know is primitive recursive. Another example of a Π_1 set is the set $\{x \mid W_x = \emptyset\}$ of indices of the empty function. This is Π_1 because

$$W_x = \emptyset \iff \forall v \forall t \text{ not } T(x, v, t),$$

and we know how to collapse $\forall\forall$ into a single $\forall$:

$$W_x = \emptyset \iff \forall s \text{ not } T(x, (s)_0, (s)_1).$$

In general, we can say that a relation is Π_1 if and only if it is the complement of a Σ_1 relation. This holds because of the principles

$$\text{not } \exists x \iff \forall x \text{ not} \qquad \text{and} \qquad \text{not } \forall x \iff \exists x \text{ not,}$$

sometimes called De Morgan's laws. (That is, saying that there does not exist a solution for x is equivalent to saying that every x fails to be a solution. And saying that not all x's have a property is equivalent to saying that there exists some counterexample x lacking the property.) Using these laws, we have for a Σ_1 relation R,

$$\vec{t} \in \overline{R} \iff \text{not } \exists x \, Q(\vec{t}, x) \text{ for computable } Q$$
$$\iff \forall x \text{ not } Q(\vec{t}, x)$$
$$\iff \forall x \, \overline{Q}(\vec{t}, x)$$

so that $\overline{R}$ is Π_1. Similarly, the complement of a Π_1 relation is Σ_1.

To use yet another Greek letter, say that R is Δ_1 if it is both Σ_1 and Π_1. Kleene's theorem tells us that R is Δ_1 if and only if it is a computable relation. For example, the set $\{x \mid W_x = \emptyset\}$ is Π_1 by the above, it is not computable by Rice's theorem, and hence, it is not Σ_1.

We now extend these ideas and define Σ_n and Π_n for each n:

Classification	Defining condition
Σ_1	$\exists x \, Q(\vec{t}, x)$
Π_1	$\forall x \, Q(\vec{t}, x)$
Σ_2	$\exists y \forall x \, Q(\vec{t}, x, y)$
Π_2	$\forall y \exists x \, Q(\vec{t}, x, y)$
Σ_3	$\exists z \forall y \exists x \, Q(\vec{t}, x, y, z)$
Π_3	$\forall z \exists y \forall x \, Q(\vec{t}, x, y, z)$

where Q is a computable relation. And so forth. (It has been estimated that the human mind cannot grasp the meaning of more than five alternating quantifiers.) We further define a relation to be Δ_n if it is both Σ_n and Π_n.

For example, the set Tot of indices of total computable functions on $\mathbb{N}$ is a Π_2 set because

$$x \in \text{Tot} \iff \forall v \exists t \, T(x, v, t),$$

where as before T is the ternary relation

$$T(x, v, t) \iff [\![x]\!](v){\downarrow} \text{ in } \le t \text{ steps},$$

which we know is primitive recursive. (More generally, $T^{(n)}$ was defined to be the $(n+2)$-ary relation

$$T^{(n)}(x, \vec{v}, t) \iff [\![x]\!]^{(n)}(\vec{v}){\downarrow} \text{ in } \le t \text{ steps}$$
$$\iff (\text{snap}^{(n)}(\vec{v}, x, t))_0 \ge \text{lh } x,$$

which we know is primitive recursive.) So the set Tot is at least within two quantifiers of computability.

We can also formulate these definitions by recursion on n. R is Σ_{n+1} if we have

$$\vec{t} \in R \iff \exists x \, Q(\vec{t}, x)$$

for all $\vec{t}$, for some Π_n relation Q. Dually, R is Π_{n+1} if we have

$$\vec{t} \in R \iff \forall x \, Q(\vec{t}, x)$$

for all $\vec{t}$, where Q is Σ_n. Starting from the concepts of Σ_1 and Π_1, we can use these clauses to characterize $\Sigma_2, \Pi_2, \Sigma_3, \ldots$.

Observation:
(a) *The complement of a Σ_n relation is Π_n.*
(b) *The complement of a Π_n relation is Σ_n.*

Proof. We use induction on n. We have already seen the argument for $n = 1$.

Suppose, as the inductive hypothesis, both (a) and (b) hold when $n = k$, and consider a Σ_{k+1} relation R

$$\vec{t} \in R \iff \exists x \, Q(\vec{t}, x),$$

where Q is Π_k. Then, we have

$$\vec{t} \in \overline{R} \iff \text{not } \exists x \, Q(\vec{t}, x)$$
$$\iff \forall x \text{ not } Q(\vec{t}, x) \quad \text{by De Morgan's laws}$$
$$\iff \forall x \, \overline{Q}(\vec{t}, x)$$

and by the inductive hypotheses, $\overline{Q}$ is Σ_k, and hence $\overline{R}$ is Π_{k+1}.

Similarly, the complement of a Π_{k+1} relation is Σ_{k+1}. So by induction, the observation holds for all $n \ge 1$. $\dashv$

For example, the set $\overline{\text{Tot}}$ of indices of nontotal functions is Σ_2:

$$e \in \overline{\text{Tot}} \iff \exists v \forall t \, \overline{T}(e, v, t),$$

Any Π_1 relation is also Σ_2 and Π_2 because we can use "vacuous" quantifiers:

$$\forall y\, Q(\vec{x}, y)\} \iff \exists z \forall y\, Q_1(\vec{x}, y, z)$$
$$\iff \forall z \exists y\, Q_2(\vec{x}, y, z),$$

where

$$Q_1(\vec{x}, y, z) \Leftrightarrow Q(\vec{x}, y) \quad \text{and} \quad Q_2(\vec{x}, y, z) \Leftrightarrow Q(\vec{x}, z).$$

Extending the use of vacuous quantifiers, we come to the following result.

Proposition:
(a) *Any Σ_n relation is also Δ_{n+1}.*
(b) *Any Π_n relation is also Δ_{n+1}.*

Thus, letting the noun Σ_k denote the collection of all Σ_k relations, we have the chains:

$$\Sigma_1 \subseteq \Sigma_2 \subseteq \Sigma_3 \subseteq \cdots$$
$$\Pi_1 \subseteq \Pi_2 \subseteq \Pi_3 \subseteq \cdots$$

We say that these chains define the *arithmetical hierarchy*. But there will be many relations that fall outside this hierarchy (i.e., some relations are not Σ_n or Π_n for any n).

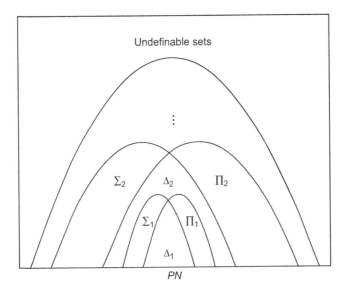

The following proposition supplies closure results under union, intersection, and substitution of total computable functions.

Proposition: *Assume that Q and R are k-ary relations on the natural numbers.*

(a) *If Q and R are both Σ_n relations, then both $Q \cup R$ and $Q \cap R$ are also Σ_n relations.*
(b) *If Q and R are both Π_n relations, then both $Q \cup R$ and $Q \cap R$ are also Π_n relations.*
    Further assume that $f_1, \ldots, f_k$ are m-place total computable functions.
(c) *If R is a Σ_n relation, then $\{\vec{x} \mid R(f_1(\vec{x}), \ldots, f_k(\vec{x}))\}$ is also a Σ_n relation.*
(d) *If R is a Π_n relation, then $\{\vec{x} \mid R(f_1(\vec{x}), \ldots, f_k(\vec{x}))\}$ is also a Π_n relation.*

Like the earlier observation concerning complements, this proposition can be verified by using induction on n. But in place of De Morgan's laws, we employ the following quantifier manipulation rules:

$$\exists x\, M(x) \text{ and } \exists y\, N(y) \iff \exists x \exists y [M(x) \text{ and } N(y)]$$
$$\exists x\, M(x) \text{ or } \exists y\, N(y) \iff \exists z [M(z) \text{ or } N(z)]$$
$$\forall x\, M(x) \text{ or } \forall y\, N(y) \iff \forall x \forall y [M(x) \text{ or } N(y)]$$
$$\forall x\, M(x) \text{ and } \forall y\, N(y) \iff \forall z [M(z) \text{ and } N(z)]$$

To see the correctness of the third rule, think about how the condition on the left side, $\forall x\, M(x)$ or $\forall y\, N(y)$, could *fail*. It fails if and only if there is both some counterexample x^* for which not $M(x^*)$ and some counterexample y^* for which not $N(y^*)$. Under what situations does the condition on the right side, $\forall x \forall y [M(x) \text{ or } N(y)]$, fail? It fails if and only if there is some counterexample $\langle x^*, y^* \rangle$ for which the condition in brackets fails so that neither $M(x^*)$ nor $N(y^*)$. And that is exactly the same situation under which the left side failed.

Proof. Parts (c) and (d) follow from known substitution rules. To prove parts (a) and (b), we use induction on n. Suppose, as the inductive hypothesis, that part (b) holds when $n = k$, and consider two Σ_{k+1} relations Q and R:

$$\vec{t} \in R \iff \exists x\, M(\vec{t}, x) \text{ where } M \text{ is } \Pi_k$$
$$\vec{t} \in Q \iff \exists y\, N(\vec{t}, y) \text{ where } N \text{ is } \Pi_k$$

Then, we have

$$\vec{t} \in R \cup Q \iff \vec{t} \in R \text{ or } \vec{t} \in Q$$
$$\iff \exists x\, M(\vec{t}, x) \text{ or } \exists y\, N(\vec{t}, y)$$
$$\iff \exists z [M(\vec{t}, z) \text{ or } N(\vec{t}, z)]$$
$$\iff \exists z [\langle \vec{t}, z \rangle \in M \cup N].$$

By the inductive hypothesis, $M \cup N$ is Π_k, and hence, $R \cup Q$ is Σ_{k+1}. A similar argument shows that $R \cap Q$ is also Σ_{k+1}. Thus part (a) holds for $n = k + 1$.

Similarly, supposing that part (a) holds when $n = k$, we find that part (b) holds for $n = k + 1$.

What about the basis for the induction? The easiest approach is to take both "Σ_0" and "Π_0" to mean simply *computable*. We know that the class of computable relations is closed under union and intersection, so parts (a) and (b) hold when $n = 0$.   ⊣

Part (c) is already familiar in the case $n = 1$; see page 89. Looking at complements will give us part (d) when $n = 1$. Then, induction on n gives us parts (c) and (d) in general.

Corollary: *Let A and B be sets of numbers with $A \leq_m B$.*

(a) If B is Σ_n, then A is also Σ_n.
(b) If B is Π_n, then A is also Π_n.

Proof. Assume that $A \leq_m B$ under f. Then $A = \{x \mid f(x) \in B\}$. Apply parts (c) and (d) of the preceding proposition.   ⊣

This corollary provides us with a method for showing that a set is *not* Σ_n or that it is *not* Π_n. We already know that to show that a set B is not r.e., one possible strategy is to try showing that $\overline{K} \leq_m B$. (Don't we?) The corollary extends the method. Whenever we have a set S that is known not to be Σ_n, then we can show that another set B is not Σ_n if we can obtain $S \leq_m B$.

But to apply this method, we first need that initial set S that is known not to be Σ_n. Read on.

For the Σ_1 relations, we have from page 82 the "normal form" result: whenever R is an n-ary Σ_1 relation, then R is the domain $W_e^{(n)}$ of some computable partial function $[\![e]\!]^{(n)}$. Hence for this e,

$$R = W_e^{(n)} = \{\vec{x} \mid \exists t\, T^{(n)}(e, \vec{x}, t)\}.$$

We want to extend this idea. First, consider a Π_1 relation R. Its complement $\overline{R}$ is Σ_1, so we can say that for some e,

$$\vec{x} \in R \iff \vec{x} \notin \overline{R}$$
$$\iff \text{not } \exists t\, T^{(n)}(e, \vec{x}, t) \quad \text{by the above}$$
$$\iff \forall t \text{ not } T^{(n)}(e, \vec{x}, t) \quad \text{by De Morgan}$$
$$\iff \forall t\, \overline{T}^{(n)}(e, \vec{x}, t).$$

We conclude that any Π_1 relation R can be written, for some number e, in the form

$$R = \{\vec{x} \mid \forall t\, \overline{T}^{(n)}(e, \vec{x}, t)\}$$

(and of course conversely any relation of this form is Π_1).

Next, consider a Π_2 relation R. We know that

$$\vec{x} \in R \iff \forall y\, Q(\vec{x}, y)$$

for some Σ_1 relation Q. Using our preceding normal form for Σ_1, we see that for some e,

$$\vec{x} \in R \iff \forall y \exists t \, T^{(n+1)}(e, \vec{x}, y, t).$$

We thus obtain the normal form for a Π_2 relation R:

$$R = \{\vec{x} \mid \forall y \exists t \, T^{(n+1)}(e, \vec{x}, y, t)\}$$

for some number e.

Keep going. For a Σ_2 relation R, we have

$$\vec{x} \in R \iff \exists y \, Q(\vec{x}, y)$$

for some Π_1 relation Q. Using our preceding normal form for Π_1, we see that for some e,

$$\vec{x} \in R \iff \exists y \forall t \, \overline{T}^{(n+1)}(e, \vec{x}, y, t).$$

For a Σ_3 relation R, we have

$$\vec{x} \in R \iff \exists z \, Q(\vec{x}, z)$$

for some Π_2 relation Q. Using our normal form for Π_2, we see that for some e,

$$\vec{x} \in R \iff \exists z \forall y \exists t \, T^{(n+2)}(e, \vec{x}, y, z, t).$$

One more. For a Π_3 relation R, we have

$$\vec{x} \in R \iff \forall z \, Q(\vec{x}, z)$$

for some Σ_2 relation Q. Using our normal form for Σ_2, we see that for some e,

$$\vec{x} \in R \iff \forall z \exists y \forall t \, \overline{T}^{(n+2)}(e, \vec{x}, y, t).$$

Let's collect what we have in a table:

Σ_1	$\{\vec{x} \mid \exists t \, T^{(n)}(e, \vec{x}, t)\}$
Π_1	$\{\vec{x} \mid \forall t \, \overline{T}^{(n)}(e, \vec{x}, t)\}$
Σ_2	$\{\vec{x} \mid \exists y \forall t \, \overline{T}^{(n+1)}(e, \vec{x}, y, t)\}$
Π_2	$\{\vec{x} \mid \forall y \exists t \, T^{(n+1)}(e, \vec{x}, y, t)\}$
Σ_3	$\{\vec{x} \mid \exists z \forall y \exists t \, T^{(n+2)}(e, \vec{x}, y, z, t)\}$
Π_3	$\{\vec{x} \mid \forall z \exists y \forall t \, \overline{T}^{(n+2)}(e, \vec{x}, y, z, t)\}$

And so forth and so on. (This table emphasizes that for any fixed k and n, there are only *countably* many Σ_n k-ary relations and only countably many Π_n k-ary relations. The last line in the table shows that each Π_3 relation has the form $\{\vec{x} \mid \forall z \exists y \forall t \, \overline{T}^{(n+2)}(e, \vec{x}, y, z, t)\}$ for some e. Taking the various possible values of e

$$e = 0, \quad e = 1, \quad e = 2, \; \ldots$$

we get a complete list, with repetitions, of all the Π_3 k-ary relations. Moreover, because a countable union of countable sets is countable, we can go a step further and say that only countably many relations can be in the arithmetical hierarchy at all. Because $\mathcal{P}\mathbb{N}$ (the power set of $\mathbb{N}$) is uncountable, there is a sense in which "most" relations fall outside the arithmetical hierarchy. See Appendix A2 for a summary of facts about countable sets.)

An advantage to having such "normal form" results is that we can diagonalize out of them. Recall that when we wanted a Σ_1 set that was not Π_1, we used the set K defined by the condition

$$x \in K \iff [\![x]\!](x) \downarrow \iff \exists t \, T(x, x, t).$$

Imitating this construction, define the Π_2 set S by the condition

$$x \in S \iff \forall y \exists t \, T^{(2)}(x, x, y, t).$$

Is it possible that this set S is also Σ_2? If so, then by our normal form results, it would have to be, for some number e, the set

$$V_e = \{x \mid \exists y \forall t \, \overline{T}^{(2)}(e, x, y, t)\}.$$

But S and V_e cannot be the same set because they differ at the number e:

$$e \notin V_e \iff \text{not } \exists y \forall t \, \overline{T}^{(2)}(e, e, y, t)$$
$$\iff \forall y \exists t \, \text{not } \overline{T}^{(2)}(e, e, y, t)$$
$$\iff \forall y \exists t \, T^{(2)}(e, e, y, t)$$
$$\iff e \in S$$

so e belongs to one and only one of the two sets S and V_e.

We conclude that S is Π_2 but not Σ_2. So its complement $\overline{S}$ is Σ_2 but not Π_2. We can generalize the construction of S to obtain the following result.

Hierarchy theorem: *For each positive integer n, there is some set that is Σ_n but not Π_n, and there is some set that is Π_n but not Σ_n.*

Example: We know that the set Tot of indices of total functions is Π_2. We can now show that it is *not* Σ_2. Take S to be the above set that is Π_2 but not Σ_2. By Exercise 3, we have $S \leq_m$ Tot. Now apply the earlier corollary: Tot cannot be Σ_2, lest S be Σ_2.

Exercises

1. **(a)** Show that $\{x \mid W_x \text{ is infinite}\}$ is Π_2.
   (b) Show that $\{x \mid \overline{W_x} \text{ is infinite}\}$ is Π_3.
2. Show that $\{x \mid W_x \text{ is a computable set}\}$ is Σ_3.
3. Show that every Π_2 set of natural numbers is many-one reducible to Tot.

   Suggestion: For a set $\{x \mid \forall u \exists v\, R(x, u, v)\}$, look at the function $\langle u, x \rangle \mapsto \mu v\, R(x, u, v)$ and apply the parameter theorem.

4. Show that the binary relation $\{\langle x, y \rangle \mid W_x \subseteq W_y\}$ is a Π_2 relation.
5. Let Z be the set of indices for the function that is constantly zero:

$$Z = \{t \mid [\![t]\!](x) = 0 \text{ for all } x\}$$

   (a) Show that Z is Π_2.
   (b) Show that Z is not Π_1.
   (c) Show that Z is not even Σ_2.

5.2 Definability in Arithmetic

A number is prime if it is greater than 1 and is not the product of two smaller numbers. That sentence reflects a certain property of the set of primes: the set of primes is *definable in arithmetic*. What other sets are definable in arithmetic? What sets are not?

Before tackling either of these questions, we need to be more explicit about what counts as "arithmetic." We want to establish a certain language so that we can then consider what is expressible in that language and what is not.

The study of definability in formal languages is an important part of logic. What we do here is to take an initial look at one such situation.

The language we want incorporates the following seven elements.

- A symbol 0 to name zero. We need to start somewhere.
- A symbol S for the successor function (that is, the function $S(x) = x + 1$). The string $S0$ names 1, the string $SS0$ names 2, and so forth. For each natural number n, we have a *numeral* $SS \cdots S0$ naming n; call this numeral $\bar{n}$. For example, $\bar{4}$ is the string $SSSS0$.
- Symbols for addition, multiplication, and exponentiation. (Everyday notation uses + and × for addition and multiplication, but lacks a symbol for exponentiation. The practice of writing x^y curiously avoids having a symbol for the exponentiation operation.)
- Symbols for comparing numbers: $=, <, \le$.
  And then some infrastructure.
- Variables $x_1, x_2, x_3, \ldots$ and $u, v, w, \ldots$. There are enough variables that we will never run out. (This is only part of the story. Actually, it is important to make the total supply of symbols *finite*. So what the language really has is one or two variables and a prime symbol, $'$. That way we can make all the variables $x', x'', x''', \ldots$ we need, with just a few basic symbols.)
- Connective words "and," "or," "not," "if ... then," and "if and only if." Also parentheses, so we don't get confused.
- Quantifiers over $\mathbb{N}$: $\forall v$ and $\exists v$ (for a variable v of our choice), to express "for every natural number" and "for some natural number."

And that is all.

Example: In the language, we can say

$$S0 < x_1 \text{ and not } \exists u \exists v(u < x_1 \text{ and } v < x_1 \text{ and } u \cdot v = x_1),$$

which expresses "x_1 is prime." If we are told what number the variable x_1 names, then we can try to say whether this expression – call it $\pi(x_1)$ – is true or false. Or better, if we *replace* the variable x_1 by a numeral $SS \cdots S0$, we get a sentence that is either true or false. That is, $\pi(SSS0)$ is true, but $\pi(SSSS0)$ is false. Or to use the abbreviations for numerals, $\pi(\bar{3})$ is true, but $\pi(\bar{4})$ is false.

Example: Fermat's Last Theorem can be written as a sentence in the language.

Nonexample: The language does not incorporate a way to say "for every *set* of natural numbers." It has no way to refer to real numbers in general or to points and lines. It talks only of natural numbers, their sums, their products, and so forth.

Definition: A set S of natural numbers is *definable in arithmetic* by an expression $\alpha(x_1)$ (of the language of arithmetic) if the following conditions hold for each number n:

(i) If $n \in S$, then $\alpha(\bar{n})$ is a true sentence.
(ii) If $n \notin S$, then $\alpha(\bar{n})$ is a false sentence.

Example: The set of primes is definable in arithmetic by the expression $\pi(x_1)$ we have just seen.

Example: The set of odd numbers is defined in arithmetic by the expression $\exists y \, x_1 = y + y + S0$. (There are several claims being made here. First, this expression is indeed in the language of arithmetic; it employs only features from our given list. Secondly, for any odd number n, the result of replacing x_1 by $\bar{n}$ is a true sentence. And thirdly, for any even number n, the result of replacing x_1 by $\bar{n}$ is a false sentence.)

Nonexample: There must be many sets that are *not* definable in arithmetic. There are uncountably many subsets of $\mathbb{N}$, by Cantor's theorem. But only countably many can be definable in arithmetic. This is because there can be only countably many defining expressions. Each expression is a finite string of symbols, drawn from a finite alphabet, and there are only countably many such strings.

The definability concept extends naturally to relations on $\mathbb{N}$.

Definition: A k-ary relation R on natural numbers is *definable in arithmetic* by an expression $\alpha(x_1, \ldots, x_k)$ (of the language of arithmetic) if the following conditions hold for each k-tuple of numbers $\langle n_1, \ldots, n_k \rangle$:

(i) If $\langle n_1, \ldots, n_k \rangle \in R$, then $\alpha(\bar{n}_1, \ldots, \bar{n}_k)$ is a true sentence.
(ii) If $\langle n_1, \ldots, n_k \rangle \notin R$, then $\alpha(\bar{n}_1, \ldots, \bar{n}_k)$ is a false sentence.

Example: The divisibility relation (which is a binary relation) is defined in arithmetic by the expression $\exists y \, x_1 \cdot y = x_2$. Call this expression $\delta(x_1, x_2)$. Then, $\delta(\bar{7}, \bar{91})$, which is the sentence $\exists y \, \bar{7} \cdot y = \bar{91}$, is true, because we can take $y = 13$.

Example: Let A be the binary relation of being "adjacent primes." That is, $\langle p, q \rangle \in A \Leftrightarrow$ both p and q are prime and $p < q$ and there is no prime in between. (For example,

$\langle 3, 7 \rangle \notin A$ and $\langle 13, 17 \rangle \in A$.) Then, A is defined in arithmetic by the expression:

$$\pi(x_1) \text{ and } \pi(x_2) \text{ and } x_1 < x_2 \text{ and not } \exists z(\pi(z) \text{ and } x_1 < z < x_2),$$

where $\pi(x_1)$ is the earlier expression defining the primes.

Our goal is to show that every relation that is Σ_n or Π_n (for any n) is definable in arithmetic. A more immediate goal is to show that the graph of every primitive recursive function is definable in arithmetic.

Example: The graph of the one-place function $f(t) = \lfloor t/2 \rfloor$ is a binary relation and is defined in arithmetic by the expression

$$x_2 + x_2 = x_1 \text{ or } x_2 + x_2 + S0 = x_1.$$

The initial functions present no difficulties:

1. The k-place function that is constantly 0 has a graph that is defined in arithmetic by the expression $x_{k+1} = 0$.
2. The successor function (which is one-place) has a graph that is defined in arithmetic by the expression $x_2 = Sx_1$.
3. The projection function I_m^k (where $1 \le m \le k$) has a graph that is defined in arithmetic by the expression $x_{k+1} = x_m$.

Now for a more serious matter.

Theorem: *The class of functions with graphs definable in arithmetic is closed under composition. That is, if f and $g_1, \ldots, g_k$ all have graphs definable in arithmetic and if h is given by the equation $h(\vec{t}) = f(g_1(\vec{t}), \ldots, g_k(\vec{t}))$, then the graph of h is also definable in arithmetic.*

Proof (for two-place functions). Assume the following:
  The graph of f is defined in arithmetic by the expression $\varphi(x_1, x_2, x_3)$.
  The graph of g_1 is defined in arithmetic by the expression $\gamma_1(x_1, x_2, x_3)$.
  The graph of g_2 is defined in arithmetic by the expression $\gamma_2(x_1, x_2, x_3)$.
And let $h(p, q) = f(g_1(p, q), g_2(p, q))$. We claim that h is defined in arithmetic by the following expression:

$$\exists y_1 \exists y_2 [\gamma_1(x_1, x_2, y_1) \text{ and } \gamma_2(x_1, x_2, y_2) \text{ and } \psi(y_1, y_2, x_3)].$$

Call this expression $\sigma(x_1, x_2, x_3)$. If $h(a, b) = c$, then $\sigma(\bar{a}, \bar{b}, \bar{c})$ is a true sentence because we can assign $g_1(a, b)$ to y_1 and $g_2(a, b)$ to y_2.

  Conversely, suppose that $\sigma(\bar{a}, \bar{b}, \bar{c})$ is a true sentence. So there must be numbers assigned to y_1 and y_2 making the expression true. The number assigned to y_1 must have been $g_1(a, b)$ to make $\gamma_1(\bar{a}, \bar{b}, y_1)$ true. Similarly, the number assigned to y_2 must have been $g_2(a, b)$ to make $\gamma_2(\bar{a}, \bar{b}, y_2)$ true. Consequently, c must be $f(g_1(a, b), g_2(a, b))$ to make $\varphi(y_1, y_2, \bar{c})$ true. ⊣

It remains to show closure under primitive recursion. Toward that end, we will employ the following two lemmas regarding two specific primitive recursive functions.

Lemma: *The graph of the function $t \mapsto p_t$ (where p_t is the $(t+1)$st prime number) is definable in arithmetic.*

Proof. We already have an expression $\delta(x_1, x_2)$ defining in arithmetic the divisibility relation and an expression $\alpha(x_1, x_2)$ defining in arithmetic the relation of being adjacent primes.

First, consider the relation Q for which

$$\langle b, c \rangle \in Q \iff b \text{ is prime and } c = 2^0 3^1 5^2 \cdots b^{\square}$$

where $\square$ is the number for which $b = p_{\square}$. For example, $\langle 5, 75 \rangle \in Q$ because $5 = p_2$ and $75 = 2^0 3^1 5^2$. Here are the first four members of Q:

$$Q = \{\langle 2, 1 \rangle, \langle 3, 3 \rangle, \langle 5, 75 \rangle, \langle 7, 25725 \rangle, \dots\}.$$

In general, we can say that $\langle b, c \rangle \in Q$ if and only if b is prime and

(i) $2 \nmid c$,
(ii) for any adjacent primes q and r with $q < r \leq b$, we have

$$q^j \mid c \iff r^{j+1} \mid c$$

for all j, and
(iii) no prime larger than b divides c.

Translating these conditions into the language of arithmetic, we obtain an expression defining Q:

> $\pi(x_1)$ and not $\delta(SS0, x_2)$ and
> $\forall u \forall v[\text{if } (\alpha(u, v) \text{ and } v \leq x_1), \text{ then } \forall w(\delta(u^w, x_2) \text{ if and only if } \delta(v^{Sw}, x_2))]$
> and not $\exists z(\pi(z) \text{ and } x_1 < z \text{ and } \delta(z, x_2))$

Call this expression $\theta(x_1, x_2)$.

Secondly, observe that $p_a = b$ if and only if b is prime, and, where c is the unique number for which $\langle b, c \rangle \in Q$, we have $b^a \mid c$ and $b^{a+1} \nmid c$. (For example, $p_2 = 5$ because 5 is prime, $5^2 \mid 75$, and $5^3 \nmid 75$.) Thus, the expression

$$\pi(x_2) \text{ and } \exists y[\theta(x_2, y) \text{ and } \delta(x_2^{x_1}, y) \text{ and not } \delta(x_2^{Sx_1}, y)]$$

defines the graph of the function $t \mapsto p_t$ in arithmetic. $\dashv$

Lemma: *The graph of the decoding function $\langle s, t \rangle \mapsto (s)_t$ is definable in arithmetic.*

Proof. The key fact is that

$$(s)_t = \begin{cases} 0 & \text{if } s = 0 \\ 0 & \text{if } p_t \nmid s \\ \text{the } e \text{ for which } p_t^{e+1} \mid s \text{ and } p_t^{e+2} \nmid s & \text{otherwise.} \end{cases}$$

Using the expression $\delta(x_1, x_2)$ for divisibility and the expression $\psi(x_1, x_2)$ for the graph of $t \mapsto p_t$, we can make the expression

$(x_1 = 0$ and $x_3 = 0)$ or
$\exists y(\psi(x_2, y)$ and not $\delta(y, x_1)$ and $x_3 = 0)$ or
$\exists y(\psi(x_2, y)$ and $\delta(y^{Sx_3}, x_1)$ and not $\delta(y^{SSx_3}, x_1))$,

which defines the graph of $\langle s, t \rangle \mapsto (s)_t$ in arithmetic. (Every number divides 0, so there is no danger that the three clauses might overlap.) ⊣

Theorem: *The class of functions with graphs definable in arithmetic is closed under primitive recursion. That is, if f and g have graphs definable in arithmetic, and if h is given by the recursion equations*

$$h(\vec{r}, 0) = f(\vec{r}) \quad \text{and} \quad h(\vec{r}, t+1) = g(h(\vec{r}, t), \vec{r}, t),$$

then the graph of h is also definable in arithmetic.

Proof. The key fact is that $h(\vec{r}, t) = q$ if and only if there exists some number s with the following three properties:

(i) $(s)_0 = f(\vec{r})$.
(ii) $(s)_{j+1} = g((s)_j, \vec{r}, j)$ for each j, unless $j \geq t$.
(iii) $(s)_t = q$.

(In one direction, if we have the equation $h(\vec{r}, t) = q$, then taking

$$s = [h(\vec{r}, 0), h(\vec{r}, 1), \ldots, h(\vec{r}, t)],$$

we see that (i), (ii), and (iii) all hold. In the other direction, suppose that s is a number satisfying (i), (ii), and (iii). Then, by induction of j, we see that $(s)_j = h(\vec{r}, j)$ for $j \leq t$. In particular, $(s)_t = q = h(\vec{r}, t)$.)

For notational simplicity, suppose that $\vec{r}$ is a single number r. We are given an expression $\varphi(x_1, x_2)$ defining the graph of f and an expression $\gamma(x_1, x_2, x_3, x_4)$ defining the graph of g. From the foregoing lemma, we have an expression $\beta(x_1, x_2, x_3)$ defining the graph of the decoding function $\langle s, t \rangle \mapsto (s)_t$. Then, the expression

$\exists z[\exists y(\varphi(x_1, y)$ and $\beta(z, 0, y))$ and
$\forall u[x_2 \leq u$ or $\exists v \exists w(\beta(z, u, v)$ and $\beta(z, Su, w)$ and $\gamma(v, x_1, u, w))]$
   and $\beta(z, x_2, x_3)]$

defines the graph of h in arithmetic. (Translation hints: The variable z will be assigned a number s meeting (i)–(iii). The variable y will be assigned $f(r)$.) ⊣

Corollary: *The graph of any primitive recursive function is definable in arithmetic.*

Corollary: *Every primitive recursive relation is definable in arithmetic.*

Proof. For any k-ary primitive recursive relation R, the graph of its characteristic function is defined in arithmetic by some expression $\rho(x_1, \ldots, x_k, x_{k+1})$. Then, R is defined in arithmetic by the expression $\rho(x_1, \ldots, x_k, S0)$.   ⊣

Corollary: *Every Σ_1 relation is definable in arithmetic.*

Proof. We showed on page 82 that any Σ_1 relation had the form $\{\vec{s} \mid \exists t\, Q(\vec{s}, t)\}$ for a primitive recursive relation Q. We know that Q is definable; add one more quantifier to define $\{\vec{s} \mid \exists t\, Q(\vec{s}, t)\}$ in arithmetic.   ⊣

Digression: Work by Martin Davis, Yuri Matiyacevich, Hilary Putnam, and Julia Robinson has shown that any Σ_1 relation is definable in arithmetic by an expression

$$\exists y_1 \cdots \exists y_k\, \theta,$$

where θ contains no quantifiers at all. And it gets even better than that; θ can actually be a polynomial equation. In particular, while θ uses multiplication and addition, it does not need exponentiation. We have included exponentiation in our language in order to simplify the proofs.

But all these corollaries are mere preliminaries for the following result.

Theorem: *Any relation that is Σ_n or Π_n (for any n) is definable in arithmetic.*

Proof. Keep adding quantifiers.   ⊣

And that is where this string of results stops. Although we will not go into the topic here, the converse to the theorem also holds: the Σ_n and Π_n relations are the *only* relations that are definable in arithmetic.

5.3   The Complexity of Truth

We now know that for any Σ_{99} set S, there is an expression $\alpha(x_1)$ of arithmetic such that

$$n \in S \iff \alpha(\bar{n}) \text{ is a true sentence.}$$

That is, any Σ_{99} set is reducible, in a sense, to the set of true sentences in arithmetic. And the same holds for any set that is Σ_{999} or elsewhere in the arithmetical hierarchy. This "reducibility" will be seen to demonstrate that the set of true sentences is a very complicated set. In particular, it will be seen that the set of true sentences is *undecidable*. It is not even semidecidable.

In order to describe the situation more precisely, we need to convert the set of true sentences of arithmetic to a set of numbers. That is, to each expression ε of arithmetic (which is a string of symbols), we can assign its *Gödel number* #ε, much as we assigned Gödel numbers to register machine programs. Of course, different expressions receive different Gödel numbers.

Moreover, for any fixed expression $\alpha(x_1)$, we expect the function

$$n \mapsto \#\alpha(\bar{n})$$

to be a computable function. (This function needs to go through the expression and replace occurrences of the variable x_1 by the numeral $\bar{n}$.)

Rather than to go into the specifics of Gödel numbering, let's take it for granted that the Gödel numbers can be assigned in such a way that $n \mapsto \#\alpha(\bar{n})$ is always a computable function. (This is not so unreasonable. Expressions of arithmetic are words over a certain finite language. We can code such words by numbers. Logic textbooks, such as the one cited in the *References*, carry out Gödel numbering and verify that substituting a numeral for a variable is a computable procedure.)

Define True to be the set of Gödel numbers of true sentences of arithmetic:

True $= \{\#\tau \mid \tau$ is a true sentence of arithmetic$\}$.

For example, the number

$$\#\forall x \forall y \forall z \forall n \left(n \leq 2 \text{ or } xyz = 0 \text{ or not } x^n + y^n = z^n \right)$$

belongs to the set True. An indication of the complexity of the set True is given by the following result.

Proposition: *For any set S that is definable in arithmetic, we have $S \leq_m$ True.*

Proof. Say S is defined in arithmetic by the expression $\alpha(x_1)$. Thus

$$n \in S \iff \#\alpha(\bar{n}) \in \text{True}$$

for each number n. Since $n \mapsto \#\alpha(\bar{n})$ is a computable function, we conclude that $S \leq_m$ True under this function.   ⊣

Corollary: *For any set S that is either Σ_n or Π_n for some n, we have $S \leq_m$ True.*

Proof. Any such set is definable in arithmetic; apply the preceding proposition.   ⊣

Tarski's theorem: *The set True is not Σ_n or Π_n for any n.*

Proof. We will show that True is not Σ_{99}. Let S be a set that is Π_{99} but *not* Σ_{99}. (We know there are such sets.) By the above corollary, we have $S \leq_m$ True. Therefore, True cannot be Σ_{99}, lest S be Σ_{99}.   ⊣

In particular, True is not a computable set. That is, *truth in arithmetic is undecidable.* We have here an unsolvability result for a problem that is not formulated in terms of computability concepts (as was the halting problem). Even for arithmetic, sometimes regarded as one of the simpler branches of mathematics, the set of true sentences is not a decidable set. (Of course, if instead of calling it "arithmetic," we call it "number theory," then its undecidability comes as less of a surprise.)

Digression: Suppose that we omit exponentiation from our language, and we apply it not to $\mathbb{N}$ but to $\mathbb{R}$. That is, now $\forall x$ means "for all real numbers x," and $\exists x$ means "there exists a real number x." Then, we get a language for talking about the real number line. For example, a subset of $\mathbb{R}$ that is a finite union of intervals (open or closed) with rational endpoints is definable in this language. Perhaps surprisingly, the set True in this situation turns out to be computable! This follows from a different theorem of Tarski, from 1939.

Moreover, in addition to showing that True is not computable, we have shown that True is not Σ_1; that is, it is not recursively enumerable. (More informally, truth in arithmetic is not semidecidable.) This fact, although lacking the full strength of Tarski's theorem, is relevant to *axiomatic theories* in arithmetic, such as might be studied in a logic course.

Imagine, then, that we want to develop an axiomatic theory for arithmetic. So we need two components. First, we need to adopt some true sentences as our *axioms*. (There is a set of sentences called the *first-order Peano axioms* that is a popular choice.) Secondly, we need to adopt rules for what is an acceptable *proof*. Here there is not so much latitude; logicians have succeeded in nailing down the concept of a proof from axioms very precisely.

But there is one additional feature we expect: the binary relation

$$\{\langle \pi, \sigma \rangle \mid \pi \text{ is a proof of } \sigma\}$$

must be a *decidable* relation. That is, it is not acceptable to take simply all true sentences as axioms because then we could not effectively tell an axiom from a nonaxiom. (The set True is not computable.) A key feature of a *proof* is that it should be *effectively verifiable*. It must be possible – in principle – for a hard-working graduate student (or a referee) to check a proof line by line and verify its correctness. We cannot demand that the student contribute brilliant insights. Nor can we demand that the student spend an infinite amount of time, checking an infinite number of cases. What we can insist on is that the student or the referee must eventually either conclude that the proof is correct or conclude that it is not yet acceptable. We need to be able to distinguish between proofs and nonproofs.

Using Church's thesis and Gödel numbers, this demand can be translated as follows: the binary relation

$$\{\langle p, s \rangle \mid p \text{ is the Gödel number of a proof of the sentence with Gödel number } s\}$$

must be a computable relation. This has the following consequence:

$$\{s \mid \exists p(p \text{ is the Gödel number of a proof of the sentence with Gödel number } s)\}$$

is recursively enumerable. (In fact, this set would be r.e. even if the binary relation, instead of being computable, were merely r.e.) That is, in an axiomatic theory of arithmetic, the set of Gödel numbers of provable sentences is recursively enumerable. Therefore, this set cannot be the same as True, which is not recursively enumerable.

The best that an axiomatic theory can hope to generate is some recursively enumerable subset T of True.

Gödel incompleteness theorem (1931): *For any recursively enumerable subset T of True, we can find a true sentence σ with $\#\sigma \notin T$.*

Thus, for an axiomatic theory of arithmetic, we can find a true sentence not provable in that theory.

First proof. T is a subset of True. These two sets cannot be equal because the first is r.e. and the second is not. So there must be something in True not in T. ⊣

But it will be more interesting if we can actually get some idea of what that sentence σ might express. Let's retrace the argument.

Second proof. We know that True is not r.e. because the non-Σ_1 set $\overline{K}$ is many-one reducible to True. That is, there is an expression $\kappa(x_1)$ that defines $\overline{K}$ in arithmetic, and

$$n \in \overline{K} \iff \#\kappa(\bar{n}) \in \text{True}.$$

For the given r.e. subset T of True, let

$$J = \{n \mid \#\kappa(\bar{n}) \in T\}.$$

Thus, J is the set of numbers that T "knows" are in $\overline{K}$. The set J is r.e. (because we have $J \leq_m T$ under the function $n \mapsto \#\kappa(\bar{n})$). So we have $J = W_j$ for some number j. Moreover, $J \subseteq \overline{K}$ (because

$$n \in J \implies \#\kappa(\bar{n}) \in T \implies \kappa(\bar{n}) \text{ is true}$$

for each n). Therefore, J is a proper subset of $\overline{K}$ because J is r.e. and $\overline{K}$ is not. So there is a number in $\overline{K}$ that is not in J. In fact, j is such a number, by the proposition on page 84. Thus, the sentence

$$\kappa(\bar{j})$$

is true (because $j \in \overline{K}$), but its Gödel number is not in T (because $j \notin J$). So here is a specific sentence witnessing Gödel's incompleteness theorem. ⊣

And what might this sentence $\kappa(\bar{j})$ say? Literally, it speaks of numbers and their sums and products – dullsville. But we can give it a more interesting translation:

$$
\begin{array}{ll}
\kappa(\bar{j}) & \text{says} \quad j \in \overline{K} \\
& \text{i.e.,} \quad j \notin W_j \\
& \text{i.e.,} \quad j \notin J \\
& \text{i.e.,} \quad \#\kappa(\bar{j}) \notin T
\end{array}
$$

That is, our witness (our true unprovable sentence) asserts, in a sense, that it is itself not in the axiomatic theory that yields T. It is saying (under this rather free translation), "I am unprovable in this axiomatic theory."

Digression: In 1931, Gödel did not have the development of computability theory available to him. Instead, he proceeded directly to an ingenious construction of a sentence that could be freely translated as saying "I am unprovable in this axiomatic theory." This sentence had to be true (if it were false, we would have a provable falsehood), and hence unprovable in the axiomatic theory. Even better, Gödel worked not from the concept of a true sentence, but from the concept (from logic) of a consistent theory. This led to a result called the *second* incompleteness theorem, which cannot be explored here.

Emil Post, in a seminal 1944 paper, defined a concept he called a *creative* set; the set K was an example of such a set. (See Exercise 9.) He gave, much as is done here, a version of the Gödel incompleteness theorem. He then added: "The conclusion is unescapable that even for such a fixed, well defined body of mathematical propositions, mathematical thinking is, and must remain, essentially creative."

Exercises

6. Call a set S of natural numbers *productive* if there is a computable partial function f (a *productive function* for S) such that whenever $W_x \subseteq S$ then $f(x)$ is defined and belongs to S but not to W_x. (Thus $f(x)$ witnesses the fact that W_x is not all of S.) For example, $\overline{K}$ is productive, and the identity function is a productive function for $\overline{K}$. Clearly, a productive set cannot be recursively enumerable. Show that if $A \leq_m B$ and A is productive, then B is also productive.

7. **(a)** Show that the set True is productive.
   (b) Show that its complement, $\overline{\text{True}}$, is also productive.

8. **(a)** Show that the set Tot is productive.
   (b) Show that its complement, $\overline{\text{Tot}}$, is also productive.

9. Call a set *creative* if it is recursively enumerable and its complement is productive. For example, the set K is creative. Show that any m-complete r.e. set (i.e., any r.e. set such that all other r.e. sets are many-one reducible to it) is creative.

6 Degrees of Unsolvability

6.1 Relative Computability

All the noncomputable sets are noncomputable, but some are more noncomputable than others. In this chapter,[1] we want to make sense of this idea.

For example, suppose A and B are both noncomputable subsets of $\mathbb{N}$. On the one hand, we might be able to show that if, hypothetically speaking, we could somehow decide membership in B, then we could decide membership in A. This would lead us to the opinion that A is no more undecidable than B is.

On the other hand, we might be able to show that even if our Fairy Godmother gave us an oracle so we could decide membership in B, there *still* would be no decision procedure for A. This might lead us to the opinion that A is *more* undecidable than B or else that their "degrees of unsolvability" are not directly comparable.

Now, an "oracle" for B sounds like a magic device – a black box sitting on a tripod, perhaps, with the ability to say whether or not a given number belongs to B. The informal concept we want to explore is the concept of effective calculability *relative to* some fixed set B. Our mental image of effective calculability (in Chapter 1) involved a clerk dutifully carrying out given instructions. Now, we intend to supply this clerk with one more asset: an oracle for B. This will allow him or her to do more than before (if B is undecidable); he or she can now very easily calculate the characteristic function C_B of B, for example. But there will still be limits to what the clerk can do, even with this new oracle.

The remarkable fact is that we can turn this into mathematics, without any magic. The concept of relative computability first appeared in a 1939 paper by Alan Turing. At first glance, it might seem strange to combine the rather constructive concept of computability with the almost mystical idea of an oracle. It is to Turing's credit that he perceived that the combination, strange or not, would be a useful tool in classifying the noncomputable sets.

The preceding paragraphs give an informal description of the concept of effective calculability relative to a set B (shades of Chapter 1!). Here is the plan. We want to make the concept into a genuine mathematical concept in two ways: general recursiveness relative to B (as in Chapter 2) and register-machine computability relative to B (as in Chapter 3). So next we will look at the (rather minor) changes needed to Chapter 2 to incorporate B. And then, we will look at the changes needed to Chapter 3 to incorporate B. In particular, we will want to know that two approaches yield exactly the same class of partial functions. And we will want to know that the theorems from

---

[1] Chapters 5, 6, and 7 are largely independent and can be read in any order.

Chapters 2 and 3 continue to hold in the new setting. After that, we will explore what use can be made of these ideas.

In Chapter 2, we formalized the concept of effective calculability by defining the class of general recursive partial functions: the partial functions that could be built up from certain initial functions by use of composition, primitive recursion, and search. Define the *general recursive partial functions relative to B* in exactly the same way, except for allowing one additional initial function, namely the characteristic function C_B of B.

Moreover, we obtain the *primitive recursive functions relative to B* by foregoing unbounded search. And an *n*-ary relation R on the natural numbers is defined to be *primitive recursive relative to B* if its characteristic function C_R is primitive recursive relative to B. Similarly, a relation R is *general recursive relative to B* if its characteristic function C_R (which is always total) is general recursive relative to B.

For a simple example, the complement $\overline{B}$ of B is primitive recursive relative to B because its characteristic function is obtainable by composition from the function $x \mapsto 1 \mathbin{\dot-} x$ and C_B.

As in Chapter 2 (and by the same proofs), the class of relations that are general recursive relative to B is closed under the constructions we came across there. In particular, it is closed under unions, intersections, complements, bounded quantifiers, and substitution of total functions that are general recursive relative to B.

(It is clear that this concept of relative general recursiveness could be extended in two ways. First, instead of adding the C_B as an initial function for *one* set B, we could add the characteristic functions of *many* sets. That is, where $\mathcal{C}$ is some collection of subsets of $\mathbb{N}$, we can define the class of general recursive partial functions relative to $\mathcal{C}$ by including all the characteristic functions of members of $\mathcal{C}$ as initial functions. And here $\mathcal{C}$ could be finite or infinite.

Secondly, for a total function F on the natural numbers, we can obtain the general recursive partial functions relative to F by adding F as an initial function. That is, there is no necessity of limiting ourselves to oracles for characteristic functions – we can handle oracles for other total functions just as well.

Combining these two extensions, where $\mathcal{F}$ is a collection of total functions, we can obtain the general recursive partial functions relative to $\mathcal{F}$ by adding all of $\mathcal{F}$ as initial functions.

These are good extensions, but we will not make use of them right now. For the time being, we will concentrate solely on the concept of computability relative to a single set B of natural numbers.)

In Chapter 3, we examined the register-machine option for formalizing the concept of effective calculability. Now, we add a new type of instruction for register machines:

- "Convert r," C r (where r is a numeral for a natural number): The effect of this instruction is to replace the number x in register r by $C_B(x)$. This happens instantly, in one step. The machine then proceeds to the next instruction in the program (if any).

(In the case of computability relative to a total function F, this instruction replaces the x in register r by $F(x)$.)

For a program $\mathcal{P}$ in the new sense, when we begin execution with $\vec{x}$ in registers $1, \ldots, n$, there are three possibilities:

(i)   The calculation might never halt.
(ii) The calculation might come to a "good" halt, by seeking the first nonexistent instruction.
(iii) The calculation might come to a "bad" halt, by seeking some other nonexistent instruction. (This can happen if the program tries to jump too far backwards or forwards; it can also happen if the last instruction is a decrement instruction.)

As in Chapter 2, we say, for an n-place partial function f, that $\mathcal{P}$ *computes f relative to B* if when we start execution with an n-tuple $\vec{x}$ in registers $1, \ldots, n$ and with 0 in all other registers, then the following conditions hold:

- If $f(\vec{x})$ is defined (i.e., if $\vec{x} \in \text{dom} f$), then the computation eventually terminates with $f(\vec{x})$ in register 0. Furthermore, the computation comes to a good halt, by seeking a $(p + 1)$st instruction, where p is the length of $\mathcal{P}$.
- If $f(\vec{x})$ is undefined (i.e., if $\vec{x} \notin \text{dom} f$), then the computation never terminates.

Then, we say that f is a partial function that is *register-machine computable relative to B* if there exists a program $\mathcal{P}$ (which may contain conversion instructions) that computes f relative to B (in the above sense of the verb "computes").

For example, the characteristic function C_B is register-machine computable relative to B: we use the program that converts register 1 and then moves the result to register 0. We can do this in four or five lines. (And those lines do not depend on what the set B is.)

All the other initial functions are register-machine computable relative to B, even without using conversion instructions. And the proofs from Chapter 3 show that the class of partial functions register-machine computable relative to B is closed under composition, primitive recursion, and search. No changes are required in any of those constructions. Therefore, we have the following result:

Theorem: *Every partial function that is general recursive relative to B is also register-machine computable relative to B.*

We can code programs as in Chapter 3, assigning Gödel numbers to conversion instructions:

$\#\mathrm{C}\, r = [4, r].$

There is no need to change the "loc" function that updates the location counter in executing a program. In the case of a conversion instruction, we already have $\text{loc}(k, m, e) = k + 1$ by the default clause.

The "mem" function for updating the memory number needs to be adapted by adding additional cases for the conversion instruction:

$$\text{mem}_B(m, c) = \begin{cases} m \cdot p_{(c)_1} & \text{if } (c)_0 = 0 \text{ and } c \neq 0 & \text{(increment)} \\ \lfloor m/p_{(c)_1} \rfloor & \text{if } (c)_0 = 1 \text{ and } p_{(c)_1} \mid m & \text{(decrement)} \\ \lfloor m/p_r^x \rfloor & \text{if } (c)_0 = 4 \text{ and } x \notin B & \text{(convert)} \\ \lfloor m/p_r^x \rfloor \cdot p_r & \text{if } (c)_0 = 4 \text{ and } x \in B & \text{(convert)} \\ m & \text{otherwise,} \end{cases}$$

where $r = (c)_1$ and $x = (m)_r^*$. In this equation, we have used only parts that are primitive recursive relative to B, so the function mem_B is primitive recursive relative to B. (Moreover, the construction tree for mem_B does not depend on what B is.) Hence, the function is register-machine computable relative to B. (And the program that computes mem_B relative to B does not depend on what B is.)

Our universal program, with mem replaced by mem_B, computes (relative to B) an $(n+1)$-place partial function $\Phi_B^{(n)}$ with the property that whenever e is the Gödel number of a program that computes (relative to B) an n-place partial function f, then

$$\Phi_B^{(n)}(e, \vec{x}) = f(\vec{x})$$

for all $\vec{x}$ (where equality of partial functions has the usual meaning). We define

$$[\![e]\!]_B^{(n)}(\vec{x}) = \Phi_B^{(n)}(e, \vec{x}).$$

Enumeration theorem:

(i) $\Phi_B^{(n)}$ *is an $(n+1)$-place partial function that is register-machine computable relative to B.*

(ii) *For each number e, the n-place partial function $[\![e]\!]_B^{(n)}$ is register-machine computable relative to B.*

(iii) *Each n-place partial function that is register-machine computable relative to B is $[\![e]\!]_B^{(n)}$ for some number e.*

Thus,

$$[\![0]\!]_B^{(n)}, \; [\![1]\!]_B^{(n)}, \; [\![2]\!]_B^{(n)}, \; \ldots$$

is a complete list (with repetitions) of all the n-place partial functions that are register-machine computable relative to B.

We obtain a snapshot function $\text{snap}_B^{(n)}$ by using mem_B in place of mem. This function is primitive recursive relative to B. And the following relation $T_B^{(n)}$ is also primitive recursive relative to B:

$$\{\langle u, \vec{x}, t \rangle \mid [\![u]\!]_B^{(n)}(\vec{x}) \downarrow \text{ in } \leq t \text{ steps}\} = \{\langle u, \vec{x}, t \rangle \mid (\text{snap}_B^{(n)}(u, \vec{x}, t))_0 \geq \text{lh } u\}.$$

Normal form theorem: *For any $\vec{x}$, e, and n,*

$$[\![e]\!]_B^{(n)}(\vec{x}) = \Phi_B^{(n)}(e, \vec{x}) = ((\text{snap}_B^{(n)}(e, \vec{x}, \mu t[(\text{snap}_B^{(n)}(e, \vec{x}, t))_0 \geq \text{lh } e]))_1)_0^*,$$

where "=" means that either both sides are undefined or else both are defined and are equal.

Looking at the right-hand side of this equation, we see that every piece defines a partial function that is general recursive relative to B. This proves that every partial function $[\![e]\!]_B^{(n)}$ that is register-machine computable relative to B is also general recursive relative to B. Combining this fact with its converse proved earlier, we have the following reassuring result.

Theorem: *A partial function is general recursive relative to B if and only if it is register-machine computable relative to B.*

Church's thesis, extended to the context of relativized computability, is the claim that not only are the two concepts in this theorem equivalent to each other, but the concepts are the *correct* concepts to formalize our ideas of effective calculability, given an oracle that can decide membership in a set B.

As in Chapter 4, we adopt a unified terminology. A partial function meeting the conditions of this theorem will be said to be a partial function *computable relative to B* (or a partial function *recursive relative to B*).

A relation on the natural numbers will be said to be *computable relative to B* (or *recursive relative to B*) if its characteristic function (which is always total) is computable relative to B.

For an extreme case, suppose $B = \emptyset$. A partial function is computable relative to $\emptyset$ if and only if it is a computable partial function (in the unrelativized sense). This holds because the characteristic function of $\emptyset$ is a constantly 0 function. This function is already included as an initial function for the construction of the general recursive partial functions. (An oracle for $\emptyset$ is not a useful device.)

Because the class of functions computable relative to B is closed under composition, we have the following:

Substitution rule: *If Q is an n-ary relation computable relative to B, and $g_1, \ldots, g_n$ are k-place total functions that are computable relative to B, then the k-ary relation*

$$\{\vec{x} \mid \langle g_1(\vec{x}), \ldots, g_n(\vec{x}) \rangle \in Q\}$$

is computable relative to B.

This concludes our review of Chapters 2 and 3, modified to allow for an "oracle" for a set B. Now, we want to make use of that material and see what we can do with it.

A key concept is that of Turing reducibility. For a subset A of $\mathbb{N}$ (that is, a unary relation), we say that A is *Turing reducible* to B (written $A \leq_T B$) if the characteristic function of A is computable relative to B. Informally, saying that $A \leq_T B$ implies that membership in A is no harder to decide than membership in B. That is, we could decide membership in A if we had an oracle for B. Of course, if A is a computable set, then automatically it is computable relative to anything one wants:

$$A \text{ computable} \implies \text{for any } B, \ A \leq_T B.$$

For example, suppose $A \leq_m B$ under the total computable function f:

$$x \in A \iff f(x) \in B.$$

Then, $C_A(x) = C_B(f(x))$ and consequently $A \leq_T B$. (Right? Method 2 is to apply composition to C_B and f. Method 3 is to follow a program for f by a conversion instruction.) That is,

$$A \leq_m B \implies A \leq_T B.$$

The converse does not hold, in general. For example, we know that $\overline{K} \not\leq_m K$ because $\overline{K}$ is not r.e. On the other hand, $\overline{K} \leq_T K$. In fact, $\overline{A} \leq_T A$ for any set A. (Right?)

Transitivity lemma: *If f is a partial function computable relative to B and $B \leq_T C$, then f is a partial function computable relative to C.*

Proof. We are given that B is general recursive relative to C, so there is construction tree $\mathcal{T}$ showing how the characteristic function of B is built up from initial functions (possibly including the characteristic function of C).

Similarly, we are given that there is a construction tree showing how the partial function f is built up. But this tree may have among its leaves the characteristic function of B.

Use grafting. In the latter tree, whenever a leaf (at the bottom of the tree) has the characteristic function of B as an initial function, we graft in the tree $\mathcal{T}$. The result is a tree that still builds up the partial function f, but now its initial functions are only zero, successor, projection, and the characteristic function of C. Hence, f is general recursive relative to C.   ⊣

Corollary: *If f is a partial function computable relative to B and B is a computable set, then f is a computable partial function (absolutely).*

Proof. In the transitivity lemma, take $C = \emptyset$.   ⊣

It is clear that $A \leq_T A$ for any set A; one says that the $\leq_T$ relation is *reflexive* on the collection $\mathcal{PN}$ of sets of natural numbers. The $\leq_T$ relation also has the following property, which is called *transitivity*.

Proposition: *Whenever $A \leq_T B$ and $B \leq_T C$, then $A \leq_T C$.*

Proof. In the transitivity lemma, take f to be the characteristic function of A.   ⊣

Proposition: *Whenever $A \leq_T B$ and B is a computable set, then A is also a computable set.*

Proof. In the previous proposition, take $C = \emptyset$.   ⊣

This proposition, in the contrapositive, has the following consequence: One strategy for showing that a set B is *not* computable is to show that $A \leq_T B$ where A is some set already known to be not computable.

Relations (such as $\leq_T$ or $\leq_m$) that are reflexive and transitive are called *preorderings*. We will examine the properties of preorderings shortly.

We say that sets A and B are *Turing equivalent* (written $A \equiv_T B$) if each is Turing reducible to the other:

$$A \equiv_T B \iff A \leq_T B \text{ and } B \leq_T A.$$

Because $\leq_T$ is reflexive on $\mathcal{PN}$, it follows that the $\equiv_T$ relation is also reflexive on $\mathcal{PN}$, that is, that $A \equiv_T A$ for every set A. Moreover, the transitivity of $\leq_T$ tells us that $\equiv_T$ is also transitive:

$$A \equiv_T B \text{ and } B \equiv_T C \implies A \equiv_T C.$$

In addition, $\equiv_T$ has a property that $\leq_T$ lacks: it is *symmetric*:

Whenever $A \equiv_T B$ then $B \equiv_T A$.

In summary, the relation $\equiv_T$ is reflexive on $\mathcal{PN}$, transitive, and symmetric. Such relations are called *equivalence relations* on $\mathcal{PN}$.

6.2 Equivalence Relations

Suppose that E is a binary relation on some set U (that is, $E \subseteq U \times U$). In place of

$$\langle x, y \rangle \in E,$$

we will write simply xEy.

Definition:

  (i) E is said to be *reflexive on U* if xEx for every x in U.
  (ii) E is said to be *symmetric* if whenever xEy then also yEx.
 (iii) E is said to be *transitive* if whenever both xEy and yEz then also xEz.
  (iv) E is said to be an *equivalence relation on U* if it is reflexive on U, symmetric, and transitive.

Example 1: The Turing equivalence relation $\equiv_T$ is an equivalence relation on $\mathcal{PN}$.

Example 2: The relation $\leq_m$ of many-one equivalence is reflexive on $\mathcal{PN}$ (that is, $A \leq_m A$) and is transitive. Define A and B to be many-one equivalent (written $A \equiv_m B$) if each is many-one reducible to the other:

$$A \equiv_m B \iff A \leq_m B \text{ and } B \leq_m A.$$

This relation is also an equivalence relation on $\mathcal{PN}$.

Example 3: For sets A and B, say that A is *one-one reducible to B* (written $A \leq_1 B$) if $A \leq_m B$ under a function that is (in addition to being total and computable) one-to-one. This relation is also reflexive on $\mathcal{PN}$ because $A \leq_1 A$ under the identity function. And it is transitive. Make the inevitable definition:

$$A \equiv_1 B \iff A \leq_1 B \text{ and } B \leq_1 A.$$

This relation is also an equivalence relation on $\mathcal{PN}$.

Example 4: Equivalence relations arise throughout mathematics, especially in algebra. For integers x and y, define $x \equiv y$ to hold if the difference $|x - y|$ is evenly divisible by 6. Then, $\equiv$ is an equivalence relation on the set $\mathbb{Z}$ of integers. If $x \equiv y$, then we say that x and y are *congruent modulo* 6.

Definition: For an equivalence relation E on a set U, and an element x in U, define the *equivalence class* $[x]_E$ of x to be the set of all objects t that x is related to:

$$[x]_E = \{t \mid xEt\}.$$

When the relation E is fixed by the context, we can write simply $[x]$.

Example 4 revisited: For congruence modulo 6 on the integers,

$$[2] = \{\ldots, -4, 2, 8, 12, \ldots\}.$$

Example 1 revisited: For Turing equivalence on $\mathcal{PN}$, the equivalence class $[\emptyset]$ of the empty set is

$$\{A \subseteq \mathbb{N} \mid \emptyset \equiv_T A\}.$$

This is exactly the collection of computable sets. On the one hand, we have $\emptyset \leq_T A$ for *any* set A because $\emptyset$ is a computable set (absolutely). On the other hand, whenever $A \leq_T \emptyset$, then A is computable, by a recent proposition.

By contrast, $[K]$ does not contain any computable sets because whenever $K \leq_T A$, then A cannot be computable (by the same proposition). Thus, $[\emptyset]$ and $[K]$ are disjoint; they have no members in common.

Lemma: *Assume that E is an equivalence relation on U and that x and y belong to U. Then,*

$$[x]_E = [y]_E \quad \text{if and only if} \quad xEy.$$

Proof.

($\Rightarrow$) Assume that $[x]_E = [y]_E$. We know that $y \in [y]_E$ (because yEy), and consequently $y \in [x]_E$ (because $[x]_E$ and $[y]_E$ are the same set). By the definition of $[x]_E$, this means that xEy.

($\Leftarrow$) Next assume that xEy. Then,

$$\begin{aligned}
t \in [y]_E &\Longrightarrow yEt \\
&\Longrightarrow xEt \quad \text{because } xEy \text{ and } E \text{ is transitive} \\
&\Longrightarrow t \in [x]_E.
\end{aligned}$$

Thus $[y]_E \subseteq [x]_E$. Since E is symmetric, we also have yEx, and we can reverse x and y in the above argument to obtain $[x]_E \subseteq [y]_E$.   ⊣

Definition: A *partition* Π of a set U is a collection of nonempty subsets of U that is disjoint and exhaustive, i.e.,

(a) no two different sets in Π have any common elements, and
(b) each element of U is in some set in Π.

There is a close connection between partitions and equivalence relations.

Theorem: *Assume that E is an equivalence relation on U. Then, the collection*

$$\{[x]_E \mid x \in U\}$$

of all equivalence classes is a partition of U.

Proof. Each equivalence class $[x]_E$ is nonempty (because $x \in [x]_E$) and is a subset of U (because $E \subseteq U \times U$). The main thing that we must prove is that the collection of equivalence classes is disjoint, i.e., part (a) of the definition is met. So suppose that $[x]_E$ and $[y]_E$ have a common element t. Thus,

xEt and yEt.

But then xEy and by the lemma, $[x]_E = [y]_E$.   ⊣

For an equivalence relation E on a set U, we can define the *quotient set*

$$U/E = \{[x]_E \mid x \in U\}$$

whose members are the equivalence classes. (The expression U/E is read "U modulo E.") This is a set of sets. We have the *natural map* (or *canonical map*) $\varphi : U \to U/E$ defined by

$$\varphi(x) = [x]_E$$

for x in U.

Example 4 revisited: For congruence modulo 6 on the integers, the quotient set $\mathbb{Z}/\equiv$ consists of exactly six sets:

$$\mathbb{Z}/\equiv = \{[0], [1], [2], [3], [4], [5]\}.$$

So $\mathbb{Z}$ has been partitioned into six parts.

Example 1 revisited: For Turing equivalence on $\mathcal{PN}$, the equivalence classes are called *degrees of unsolvability*, or *Turing degrees*, or simply *degrees*. For example, one of the degrees is $[\emptyset]$, the collection of computable sets.

The concepts of countable and uncountable sets (see Appendix A2) can be usefully applied here. The set of all possible register-machine programs is countable. One way to see this is to note that the map from programs to their indices is one-to-one. So we have a one-to-one function from the set of all programs into $\mathbb{N}$.

Proposition: *For any fixed set B, the set*

$$\mathcal{L}_B = \{A \mid A \leq_T B\}$$

is countable.

Proof. We can map each A in this set to the smallest number that is the Gödel number of a program computing A relative to B. This gives a one-to-one map from $\mathcal{L}_B$ into $\mathbb{N}$.   ⊣

Corollary: *Each Turing degree is a countable collection of sets.*

Proof. $[B]$ is a subset of $\mathcal{L}_B$.   ⊣

Proposition: *The set $\mathcal{D}$ of all Turing degrees is uncountable.*

Proof. We use the fact that the union of countably many countable sets is countable. This implies that the union of countably many degrees must be a countable subset of $\mathcal{PN}$. But the union of *all* degrees is $\mathcal{PN}$, which is uncountable.   ⊣

6.3   Preordering Relations

Suppose that R is any binary relation on a set U (that is, $R \subseteq U \times U$). As before, we can write xRy to mean that $\langle x, y \rangle \in R$.

Definition: R is a *preordering* on U if it is reflexive on U and is transitive.

Example 1: Turing reducibility $\leq_T$ is a preordering on $\mathcal{PN}$.

Example 2: Many-one reducibility $\leq_m$ is a preordering on $\mathcal{PN}$.

Example 3: One-one reducibility $\leq_1$ is a preordering on $\mathcal{PN}$.

Example 4′: Let U be the set of nonzero integers (positive or negative), and that

$$mRn \iff m \text{ divides } n.$$

Then, R is obviously reflexive on U, and transitivity is easy to check.

We want to establish that a preordering relation R on U (a) determines an equivalence relation E on U and (b) partially orders the set U/E of equivalence classes.

We obtain the equivalence relation E from the preordering R by making it symmetric. Define relation E on U by the condition:

$$xEy \iff xRy \text{ and } yRx$$

for x and y in U. (In other words, the relation E is $R \cap R^{-1}$.)

Proposition: *The relation E is an equivalence relation on U.*

Proof. The definition makes it clear that E is symmetric. It inherits reflexivity and transitivity from R: For x in U, we have xEx because xRx. If xEy and yEz, then we have four facts: xRy, yRx, yRz, and zRy. Regrouping these and using the transitivity of R, we get xRz and zRx, whence xEz.   ⊣

Hence, the relation E partitions U into equivalence classes. Let $[x]$ denote the equivalence class to which x belongs. We know that $[x] = [y]$ if and only if xEy. Now consider the quotient set U/E of all equivalence classes. We can define a binary relation $\leq$ on U/E by the condition:

$$[x] \leq [y] \iff xRy$$

for x and y in U.

Caution: There is something to prove here, namely that $\leq$ is "well defined" or "invariant." Suppose that a and b are two equivalence classes. We are attempting to define whether or not $a \leq b$ holds by *choosing* a particular x from a and a y from b, and then testing to see if xRy. We need to verify that the verdict is independent of the particular choices made. Suppose that instead of x and y, we had chosen $x' \in \boldsymbol{a}$ and $y' \in \boldsymbol{b}$. What must be shown is that $xRy \Leftrightarrow x'Ry'$.

Once we see what must be shown, actually showing it is easy. Since x and x' are in the same equivalence class, we have xEx'. Similarly yEy'. It follows from transitivity that

$$xEx' \ \& \ yEy' \text{ and } xRy \implies x'Ry'.$$

Proposition: *The relation $\leq$ is reflexive on U/E, transitive, and antisymmetric.*

"Antisymmetric" means that whenever both $\boldsymbol{a} \leq \boldsymbol{b}$ and $\boldsymbol{b} \leq \boldsymbol{a}$ then $\boldsymbol{a} = \boldsymbol{b}$. A relation that is reflexive on U/E, transitive, and antisymmetric is called a *partial ordering* on U/E.

Proof. Reflexivity and transitivity are inherited from R. Suppose that both $[x] \leq [y]$ and $[y] \leq [x]$. Then, we have both xRy and yRx, whence xEy. Therefore $[x] = [y]$. $\dashv$

For equivalence classes $\boldsymbol{a}$ and $\boldsymbol{b}$, we write $\boldsymbol{a} < \boldsymbol{b}$ to mean that $\boldsymbol{a} \leq \boldsymbol{b}$ and $\boldsymbol{a} \neq \boldsymbol{b}$.

Proposition: *The relation $<$ is irreflexive (i.e., never $\boldsymbol{a} < \boldsymbol{a}$) and transitive.*

Proof. Irreflexivity is clear. Suppose that $\boldsymbol{a} < \boldsymbol{b} < \boldsymbol{c}$. Then clearly $\boldsymbol{a} \leq \boldsymbol{c}$, but could we have $\boldsymbol{a} = \boldsymbol{c}$? No, that would imply $\boldsymbol{a} \leq \boldsymbol{b} \leq \boldsymbol{a}$, whence $\boldsymbol{a} = \boldsymbol{b}$ by antisymmetry. $\dashv$

Example 4' revisited: In Example 4' above, we have mEn if and only if $|m| = |n|$. Each equivalence class contains exactly two numbers; $[n] = \{n, -n\}$. Under the partial order, $[1]$ is the *least* class, that is, $[1] \leq [n]$ for every n. The partial order is not a total order; for example, $[2]$ and $[3]$ are incomparable. (That means that neither $[2] \leq [3]$ nor $[3] \leq [2]$.) For any class, there is a strictly larger one.

Example 1 revisited: From the preordering $\leq_T$ on $\mathcal{PN}$, we obtain the equivalence relation E which is nothing but the Turing equivalence relation $\equiv_T$. The equivalence classes are the Turing degrees, and $\leq_T$ determines a partial ordering on the Turing degrees:

$$[A] \leq [B] \iff A \leq_T B.$$

We now want to examine in more detail this partial ordering of the Turing degrees.

6.4   Ordering Degrees

The degree $[\varnothing]$ (call this degree $\boldsymbol{0}$) consisting of the computable sets is the *least* degree in this partial ordering. That is, for any degree $\boldsymbol{a}$, we have $\boldsymbol{0} \leq \boldsymbol{a}$ because $\varnothing \leq_T A$ for any set A.

Let $\boldsymbol{0}'$ be the degree of K. Then $\boldsymbol{0} < \boldsymbol{0}'$.

Define a degree to be *recursively enumerable* if it contains an r.e. set. The degree $\boldsymbol{0}$ is a recursively enumerable degree; in fact *all* of its members, being computable, are r.e. And the degree $\boldsymbol{0}'$ is a recursively enumerable degree because it contains K. (It also contains sets that are *not* recursively enumerable, such as $\overline{K}$.)

Proposition: $\mathbf{0}'$ *is the largest recursively enumerable degree. That is,* $\mathbf{a} \leq \mathbf{0}'$ *for any recursively enumerable degree* $\mathbf{a}$.

Proof. Take an r.e. set A in $\mathbf{a}$. We saw in Chapter 4 that $A \leq_m K$ because K is a complete r.e. set. Therefore $A \leq_T K$. So $\mathbf{a} \leq \mathbf{0}'$.   ⊣

In 1944, Emil Post raised the question whether there were any r.e. degrees other than $\mathbf{0}$ and $\mathbf{0}'$. This question, which became known as "Post's problem," was finally answered in 1956 (two years after Post's death), independently by Richard Friedberg (in his Harvard senior thesis) and by Albert A. Muchnik in Russia. They showed that intermediate r.e. degrees do indeed exist, and in great profusion. (Nonetheless, there seem to be no "natural" examples of such degrees.)

There is no largest degree. We will see shortly how to construct, for each degree $\mathbf{a}$, a strictly larger degree $\mathbf{a}'$.

Exercises

1. Define $A \leq_* B$ to be like $A \leq_m B$ except not requiring that the function be computable. That is, $A \leq_* B$ if there exists some total function f, computable or not, for which

   $$x \in A \iff f(x) \in B$$

   for all x. Then, $\leq_*$ is a preordering on $\mathcal{PN}$. So it determines an equivalence relation $\equiv_*$ and a partial ordering on the quotient set. How many equivalence classes are there, and how are they ordered?
2. Let U be the collection of computable subsets of $\mathbb{N}$. Then, $\leq_m$ (restricted to U) is a preordering on U. So it determines an equivalence relation $\equiv_m$ on U and a partial ordering on the quotient set. How many equivalence classes are there, and how are they ordered?
3. Give an example of a set in the degree $\mathbf{0}'$ that is neither recursively enumerable nor the complement of a recursively enumerable set.
4. Assume that A and B are sets for which $A \equiv_m B$. Show that we also have $A \equiv_T B$. (One says that the equivalence relation $\equiv_m$ *refines* the equivalence relation $\equiv_T$.)
5. Show that for any partial function f, there is some set B such that f is a partial function computable relative to B.

6.5   Structure of the Degrees

Consider any two sets, say A (of degree $\mathbf{a}$) and B (of degree $\mathbf{b}$). Then, there are the four disjoint possibilities:

- $\mathbf{a} = \mathbf{b}$. This happens when $A \equiv_T B$. We can think of A and B as being, in some sense, "equally complex," or as having the same "information content."
- $\mathbf{a} < \mathbf{b}$. This happens when $A \leq_T B$ but $A \not\equiv_T B$. We can think of A as being, in some sense, "simpler" than B is, or of having less information content than B has.

- $b < a$. We can think of B as being, in some sense, "simpler" than A is.
- None of the above. That is, a and b are *incomparable*. This happens when $A \not\leq_T B$ and $B \not\leq_T A$. Incomparable degrees do indeed exist.

Theorem: *Any two degrees have a least upper bound. That is, for any degrees a and b, we can find a degree c with the two properties*

 (i) $a \leq c$ *and* $b \leq c$, *and*
(ii) *whenever d is a degree for which both $a \leq d$ and $b \leq d$, then $c \leq d$.*

If either $a \leq b$ or $b \leq a$, this theorem is trivial; we simply take c to be the larger degree. The theorem is interesting only in the case where a and b are incomparable.

Proof. Choose a set A from degree a and a set B from degree b. Define the set

$$A \oplus B = \{2x \mid x \in A\} \cup \{2x + 1 \mid x \in B\}.$$

This set codes A on the even numbers and codes B on the odds. Let c be the degree of $A \oplus B$.

Then, $a \leq c$ because $A \leq_1 A \oplus B$ under the function $x \mapsto 2x$. Similarly, $b \leq c$ because $B \leq_1 A \oplus B$ under the function $x \mapsto 2x + 1$.

Now suppose that d is a degree with both $a \leq d$ and $b \leq d$. Choose a set D from the degree d. Then, $A \leq_T D$ and $B \leq_T D$. We seek to show that $A \oplus B \leq_T D$. The following program computes the characteristic function of $A \oplus B$ relative to D:

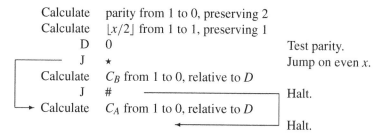

Calculate	parity from 1 to 0, preserving 2	
Calculate	$\lfloor x/2 \rfloor$ from 1 to 1, preserving 1	
D	0	Test parity.
J	⋆	Jump on even x.
Calculate	C_B from 1 to 0, relative to D	
J	#	Halt.
Calculate	C_A from 1 to 0, relative to D	
		Halt.

Thus $c \leq d$.   ⊣

A set with a partial ordering is called a *lattice* if any two elements always have a least upper bound and a greatest lower bound. The degrees do *not* form a lattice, because greatest lower bounds do not always exist. The best we can say is that we have an "upper semilattice" because we can at least take least upper bounds.

Proposition: *For any set B of natural numbers, the collection*

$$\mathcal{U}_B = \{C \mid B \leq_T C\}$$

is uncountable.

Proof. For any A in $\mathcal{PN}$, the set $A \oplus B$ is in $\mathcal{U}_B$. The map $A \mapsto A \oplus B$ is one-to-one, so $\mathcal{PN}$ has the same size as a subset of $\mathcal{U}_B$.   ⊣

Corollary: *For any degree b, the set $\{c \mid b \le c\}$ of larger degrees is uncountable.*

Proof. Look at the union of this set.

$$\bigcup \{c \mid b \le c\} = \mathcal{U}_B.$$

On the one hand, the preceding proposition tells us that $\mathcal{U}_B$ is uncountable. On the other hand, the union of countably many countable sets is countable. Hence, the union of any countable collection of degrees must be countable. ⊣

A much weaker statement is that the set $\{c \mid b \le c\}$ (sometimes called the "cone above b") contains more than one degree. So there can be no largest degree; for any degree b, there are many degrees in the cone above it.

Contrast this result with the following.

Proposition: *For any degree b, the set $\{a \mid a \le b\}$ of smaller degrees is countable.*

Proof. Fix some B in b. Map each smaller degree a to the smallest number that is the Gödel number of a register-machine program that computes relative to B some set of degree a. (We have seen this argument before, back on page 129.) ⊣

Parameter theorem: *For each m and n, there is an $(n + 1)$-place primitive recursive function ρ_{mn} such that the equation*

$$[\![e]\!]_B^{(m+n)}(\vec{x}, \vec{y}) = [\![\rho_{mn}(e, \vec{y})]\!]_B^{(m)}(\vec{x})$$

for all B, e, $\vec{x}$, and $\vec{y}$. (Here equality has the usual meaning: either both sides are undefined, or both sides are defined and are the same.) Moreover, ρ_{mn} is one-to-one.

Proof. We proceed as before. Here, $\vec{x}$ is $\langle x_1, \ldots, x_m \rangle$ and $\vec{y}$ is $\langle y_1, \ldots, y_n \rangle$.

$$\rho_{mn}(e, y_1, \ldots, y_n) = k_{m+1}(y_1) * \cdots * k_{m+n}(y_n) * k_{m+n+1}(e) * p,$$

where p is the Gödel number of our program that computes $\Phi_B^{(m+n)}$. (That program did not depend on what the set B is. Here, m, n, and p are fixed; e, $\vec{x}$, and $\vec{y}$ are the variables.) ⊣

It should be noted here that ρ_{mn} is primitive recursive absolutely, and not merely primitive recursive relative to B.

Definition: A relation on $\mathbb{N}$ is said to be *recursively enumerable relative to B* (where $B \subseteq \mathbb{N}$) if it is the domain of some partial function that is computable relative to B.

Define W_e^B to be the domain of $[\![e]\!]_B$. Then,

$$W_0^B, \quad W_1^B, \quad W_2^B, \quad \ldots$$

is a complete list with repetitions of the sets (of natural numbers) that are recursively enumerable relative to B.

As with the unrelativized concept, there are several equivalent formulations of the definition of recursive enumerability relative to an oracle. For now, here is one of them.

Theorem: *A relation R on $\mathbb{N}$ is recursively enumerable relative to B if and only if it is Σ_1 relative to B:*

$$\vec{x} \in R \iff \exists y \, Q(\vec{x}, y)$$

for some relation Q that is computable relative to B.

Proof. In one direction, suppose we have $\vec{x} \in R \iff \exists y \, Q(\vec{x}, y)$. Define the partial function:

$$f(\vec{x}) = \mu y \, Q(\vec{x}, y).$$

Then $R = \operatorname{dom} f$. And f is a partial function that is computable relative to B.
  In the other direction, suppose that $R = \operatorname{dom} [\![e]\!]_B^{(n)}$. Then,

$$\vec{x} \in R \iff \exists t \, [\, [\![e]\!]_B^{(n)}(\vec{x}) \downarrow \text{ in } \le t \text{ steps}]$$
$$\iff \exists t \, [(\operatorname{snap}_B^{(n)}(\vec{x}, e, t))_0 \ge \operatorname{lh} e],$$

which is Σ_1 relative to B. $\dashv$

Proposition: *Whenever A is recursively enumerable relative to B and $B \le_T C$, then A is recursively enumerable relative to C.*

Proof. Apply the transitivity lemma to a partial function that has domain A and is computable relative to B. $\dashv$

For the recursively enumerable relations, our standard examples in Chapter 4 were the halting relation and the set K. The definition of K can be relativized.

Definition: For a set B of natural numbers, its *jump* is the set

$$B' = \{x \mid [\![x]\!]_B(x) \downarrow\}.$$

Theorem:

(a) *The set B' is recursively enumerable relative to B.*
(b) *B' is not computable relative to B. In fact its complement is not even recursively enumerable relative to B.*

Proof.

(a) B' is the domain of the function $x \mapsto [\![x]\!]_B(x)$ which by the enumeration theorem is a partial function computable relative to B.

(b) If B' were computable relative to B, then its complement would also be computable relative to B and hence recursively enumerable relative to B. So it suffices to prove the second half of part (b).

Any set that is recursively enumerable relative to B must be W_e^B for some e. But by the definition of B', we have

$$e \in \overline{B'} \iff e \notin W_e^B.$$

So $\overline{B'} \neq W_e^B$ because they differ at e.   ⊣

Proposition: *If $A \leq_m C$ and C is recursively enumerable relative to B, then A is also recursively enumerable relative to B.*

Proof. Say f is a total computable function that many-one reduces A to C. Because C is Σ_1 relative to B, we have, for some relation Q that is computable relative to B,

$$x \in A \iff f(x) \in C$$
$$\iff \exists y \, Q(f(x), y)$$

and the relation $\{\langle x, y \rangle \mid Q(f(x), y)\}$ is computable relative to B.   ⊣

Theorem: *For sets A and B of natural numbers,*

$$A \text{ is recursively enumerable relative to } B \iff A \leq_1 B'.$$

Proof. The preceding proposition covers the "$\Leftarrow$" direction (where we take C to be B'). The "$\Rightarrow$" direction asserts that B' is a 1-complete recursively enumerable set relative to B.

So assume that A is recursively enumerable relative to B, say $A = W_a^B$. Our plan is to make a one-to-one primitive recursive function g such that for each y,

$$y \in A \Rightarrow [\![g(y)]\!]_B \text{ is total} \Rightarrow g(y) \in B'$$
$$y \notin A \Rightarrow [\![g(y)]\!]_B \text{ is empty} \Rightarrow g(y) \notin B'.$$

Consider the two-place function $\langle x, y \rangle \mapsto [\![a]\!]_B(y)$ (which is a partial function $[\![e]\!]_B^{(2)}$ computable relative to B) and parameterize out y. Then, taking $g(y) = \rho(e, y)$, we have the condition we sought.   ⊣

In particular, taking $A = B$ in this theorem, we see that $B \leq_1 B'$. Putting this together with the fact that $B' \not\leq_T B$, we conclude that $[B] < [B']$. That is, the jump operation strictly increases the degree of a set.

Lemma: *Whenever $A \leq_T B$, then $A' \leq_1 B'$.*

Proof. We assume that $A \leq_T B$. We know that A' is recursively enumerable relative to A. So by a recent proposition, A' is also recursively enumerable relative to B. So by the preceding theorem, $A' \leq_1 B'$.   ⊣

Applying this lemma twice, we see that whenever $A \equiv_T B$, then $A' \equiv_1 B'$, and consequently $A' \equiv_T B'$. This fact allows us to make a jump operation *on degrees*.

Definition: For a degree a, define its *jump* a' to be the degree $[A']$, where A is any set chosen from the degree a. (The preceding lemma tells us that the degree a' does not depend on which set A is chosen from a.)

Because $[A] < [A']$, we can conclude that on the degrees,

$$a < a' < a'' < a''' < \cdots$$

for any degree a. Again, we see that there is no largest degree.

Earlier, we defined $0'$ to be the degree of K. Now, we are saying that $0'$ is the degree of the jump of a computable set. There is no conflict here; the degrees are the same.

The converse of the preceding lemma also holds.

Proposition: *Whenever* $A' \leq_m B'$, *then* $A \leq_T B$.

Proof. On the one hand, we have $A \leq_1 A' \leq_m B'$, so A is recursively enumerable relative to B.

On the other hand, $\overline{A}$ is recursively enumerable relative to A, and so $\overline{A} \leq_1 A' \leq_m B'$.

Hence, both A and $\overline{A}$ are recursively enumerable relative to B. By Kleene's theorem (converted to relativized form), $A \leq_T B$.   ⊣

The jump operation can be thought of as providing us with a unit of measurement in the degrees. That is, a set of degree a''' is "three jumps" more complicated than a set of degree a. In particular, a set of degree $0'''$ is three jumps away from computability.

There is also a connection between relative computability and the arithmetical hierarchy, as in Chapter 5. The following result, here stated without proof, extends the fact that a relation is Σ_1 if and only if it is r.e.

Post's theorem:

(a) A relation is Σ_2 if and only if it is r.e. relative to $\emptyset'$, the jump of the empty set.
(b) More generally, a relation is Σ_{k+1} if and only if it is r.e. relative to $\emptyset^{(k)}$, the kth jump of the empty set.

Exercises

6. Let $H = \{2^{u+1}3^{v+1} \mid [\![u]\!](v) \downarrow\}$. (Thus, H encodes the halting problem.) Show that the partial function

    $$f(x) = \text{the least member (if any) of } W_x$$

    is computable relative to H.
7. Let $H = \{2^{u+1}3^{v+1} \mid [\![u]\!](v) \downarrow\}$. (Thus, H encodes the halting problem.) Show that the partial function

    $$g(x) = \text{the least member (if any) of } \overline{W_x}$$

    is computable relative to H.

8. Let Fin $= \{x \mid W_x$ is finite$\}$. Let

$$B = \{2^{x+1}3^{y+1} \mid W_x \text{ contains a number } \geq y\}.$$

Show that Fin is recursively enumerable relative to B.

9. Let $R = \{x \mid W_x$ is a computable set$\}$. Let

$$C = \{2^{u+1}3^{v+1} \mid W_u \text{ and } W_v \text{ are complementary}\}.$$

Show that R is recursively enumerable relative to C.

10. Show that $\overline{\text{Tot}}$ is recursively enumerable relative to K.

11. Give an example of sets A, B, and C for which A is recursively enumerable relative to B, B is recursively enumerable relative to C, but A is *not* recursively enumerable relative to C. (The point of this exercise is to show that relative recursive enumerability is *not* transitive.)

12. Call a set S of degrees *large* if it includes, for some degree a, the entire "cone" $\{c \mid a \leq c\}$. Show that the intersection of any two large sets of degrees is again large.

13. For each of the following sets of degrees, determine whether or not it is large, in the sense of the preceding exercise. Also determine whether or not its complement is large.

   (a) The set of recursively enumerable degrees.

   (b) The set of degrees a such that every set in a is infinite.

14. Let $H = \{2^{u+1}3^{v+1} \mid [\![u]\!](v) \downarrow\}$. (Thus, H encodes the halting problem.) Show that there is a total function f computable relative to H that dominates every total computable function g (in the sense that $f(x) > g(x)$ for all but finitely many values of x).

7 Polynomial-Time Computability

7.1 Feasible Computability

Up to now, we have approached computability from the point of view that there should be no constraints either on the time required for a particular computation or on the amount of memory space that might be required. The result is that some total computable functions will take a very long time to compute. If a function f grows very rapidly, then for large x, it will take a long time simply to generate the output $f(x)$. But there are also bounded functions that require a large amount of time.

In this chapter,[1] we want to adopt a different attitude, and we want to pay some attention to the running time that a program might require. For the most part, we will not attempt to give complete rigorous proofs. Instead, we will concentrate on introducing the concepts and describing the ideas behind the results, and on giving pointers to topics for further study.

Of course, the complexity of computing $f(x)$ depends on the program used. Suppose we define $M(e, x)$ to be the number of steps the program with Gödel number e uses on input x:

$$M(e, x) = \mu t\, T(e, x, t)$$
$$= \mu t([\![e]\!](x) \downarrow \text{ in } \leq t \text{ steps})$$

Then the computable partial function M has the following two properties:

- $M(e, x) \downarrow \iff [\![e]\!](x) \downarrow$.
- The ternary relation $\{\langle e, x, t \rangle \mid M(e, x) \downarrow \text{ and } M(e, x) \leq t\}$ is a computable relation.

In fact, that ternary relation is nothing but the primitive recursive relation $T(e, x, t)$. It can be thought of more simply as $\{\langle e, x, t \rangle \mid M(e, x) \leq t\}$ if we adopt the convention that whenever $M(e, x) \uparrow$, then $M(e, x) = \infty$ and $\infty > t$, for any t. (That is, computations that never halt take "infinite time.")

Digression: Michael Rabin and Manuel Blum have developed a theory of "axiomatic complexity" based on having a computable partial function M meeting the two conditions listed earlier. Such a function M is called a *measure of complexity*. Time (i.e., the number of steps) is a measure of complexity, but so is space. That is, if we define $M^*(e, x)$ to be the sum of the largest number of symbols that get put into the registers (this being a measure of the "space" program e uses on input x) whenever $[\![e]\!](x) \downarrow$, then M^* also meets the conditions to be a measure of complexity.

---

[1] Chapters 5, 6, and 7 are largely independent and can be read in any order.

Computability Theory
Copyright © 2011 Elsevier Inc. All rights reserved.

The following theorem asserts that we can find subsets of $\mathbb{N}$ whose decision problems are arbitrarily difficult. For example, suppose we take some function that grows rapidly, such as $h(x) = 2^{2^x}$. The theorem provides a subset A of $\mathbb{N}$ that, on the one hand, is computable (so there *are* programs that decide membership in A), and on the other hand, it has the feature that *any* program that decides membership in A will need, on input x, at least 2^{2^x} steps, with the exception of finitely many inputs. (Some programs might have a built-in table for the first thousand members of A, so that the programs are very fast for small values of x. But from some point on, the programs will be slow.)

Rabin's theorem (1960): *For any total computable function h, we can find a total computable function $f : \mathbb{N} \to \{0, 1\}$, such that for any index i of f,*

$M(i, x) > h(x)$ *except for finitely many values of x.*

Proof outline. We will construct f in such a way that whenever some number j is an index for a fast function, then f differs from that function.

In calculating $f(x)$ for a particular x, we might "eliminate" some number j. (Eliminating j will assure that it cannot be an index of f.) So suppose we already know what numbers, if any, have been eliminated in computing $f(0), \ldots, f(x-1)$.

To compute $f(x)$, we first find the least $j \leq x$ (if any) such that

- j is fast at x, that is, $M(j, x) \leq h(x)$
- j has not already been eliminated in computing $f(0), \ldots, f(x-1)$.

If there is no such j, then let $f(x) = 0$. But if we find such a j, then we eliminate it, and we let $f(x) = 1 \dot{-} [\![j]\!](x)$. (This assures that j cannot be an index for f because $[\![j]\!]$ and f differ at x.)

That completes the construction of f. Does it work? At least $f : \mathbb{N} \to \{0, 1\}$ and f is total and computable. Now let i be any index of f. Then i was never eliminated. Why not? Take any x large enough that $x \geq i$ and anything below i that was ever going to be eliminated has been eliminated before we come to $f(x)$. Then in computing $f(x)$, we would have eliminated i if $M(i, x) \leq h(x)$ had held. So it did not hold; that is, $M(i, x) > h(x)$. $\dashv$

The conclusion of Rabin's theorem, that some computable functions are really hard to compute, conforms to our informal feeling about such matters. What is more unexpected is that the feeling can be formulated as a precise theorem.

Is there a more restricted concept of "feasibly computable function" where the amount of time required does not grow beyond all reason and is an amount that might actually be practical, at least when the input is not absurdly large? To this very vague question, an exact answer has been proposed.

Definition: Call a function f *polynomial-time computable* (or for short, P-time computable) if there exists a program e for f and a polynomial p such that for every x, the program e computes $f(x)$ in no more than $p(|x|)$ steps, where $|x|$ is the length of x.

This definition requires some explanation and support. If f is a function over Σ^*, the set of words over some finite alphabet Σ, then of course $|x|$ is just the number of symbols in the word x. If f is a function over $\mathbb{N}$, then $|x|$ is the length of the *numeral* for x. (Here we come again to the fact that effective procedures work with numerals, not numbers.) So if we use base-2 numerals for $\mathbb{N}$ (either binary or dyadic notation), then $|x|$ is about $\log_2 x$.

Moreover, we were vague about how the number of steps in a computation was to be determined. Here the situation is very encouraging: The class of P-time computable functions is the same, under the different reasonable ways of counting steps. For definiteness, we adopt the following conventions: The computing is to be done by register machines operating on words over the alphabet $\{0, 1\}$, and the numerals are to be binary numerals. Thus, for example, in computing $f(2055)$, we are allowed at most $p(12)$ steps because the binary numeral for 2055 has 12 bits. (In particular, it should be emphasized that we are *not* allowed $p(2055)$ steps. If our program is trying to find the smallest prime factor of a 900-bit number, then we are allowed $p(900)$ steps, and not $p(2^{900})$ steps.)

The "industry standard" conventions are to use Turing machines and binary numerals. Here we have opted for register machines to make use of the material already developed for register machines.

Earlier, we came across the encouraging fact that many different ways of formalizing the concept of effective calculability yielded exactly the same class of functions. For example, the class of functions computable by Turing machines operating on words over $\{0, 1\}$ coincides with the class of functions computable by register machines operating on words over $\{0, 1\}$. As remarkable as that fact is, even more is true. The number of steps required in one case is bounded by a polynomial in the number of steps required by the other. For example, there exists a polynomial p (of moderate degree) such that a computation by a Turing machine that requires n steps can be simulated by a register machine that requires not more than $p(n)$ steps. Consequently, the concept of a P-time computable function is robust: We get the same class of functions, regardless of which choice we make. To be sure, the degrees of the polynomials will vary somewhat, but the class of P-time functions is unchanged. Moreover, this equivalence extends to other "reasonable" formalizations of computability.

Encouraged by this result, and inspired in particular by the 1971 work of Stephen Cook, people since the 1970s have come to regard the class of P-time functions as the correct formalization of the idea of functions for which computations are feasible, without totally impractical running times.

By analogy to Church's thesis, the statement that P-time computability corresponds to feasibly practical computability has come to be known as *Cook's thesis* or the *Cook–Karp thesis*. (The concept of P-time computability appeared as early as 1964 in the work of Alan Cobham. Jack Edmunds, in 1965, pointed out the good features of P-time algorithms. Richard Karp, in 1972, extended Cook's work.)

The following table gives a small illustration. Suppose the input string consists of n bits. If we can execute a million steps per second, then the time required to execute

	$n = 20$	$n = 30$	$n = 40$	$n = 50$	$n = 60$
$q(n) = n^3$	< 1 sec.	< 1 sec.	< 1 sec.	< 1 sec.	< 1 sec.
$q(n) = n^5$	3.2 sec.	24.3 sec.	1.7 min.	5.2 min.	13.0 min.
$q(n) = 2^n$	1.0 sec.	17.9 min.	12.7 days	35.7 years	366 centuries
$q(n) = 3^n$	58 min.	6.5 years	3,855 centuries	2×10^8 centuries	the age of the universe

$q(n)$ steps is, of course, $q(n)$ microseconds. The table converts $q(n)$ microseconds to more comprehensible units, for five choices of n and four choices of q.

So what are the P-time computable functions? As a lower bound, we can show that all of the polynomial functions are P-time computable, as are some functions that grow faster than any polynomial.

But first, we will look at upper bounds. For a start, we can say that the P-time computable functions form a subclass of the primitive recursive functions.

Theorem: *Any P-time computable function is primitive recursive.*

Proof idea. We saw earlier (in Exercise 6, page 71) that any function computable in primitive recursive time is primitive recursive. The argument there can be adapted to the present situation, where we are using register machines over a two-letter alphabet.

Suppose we have a program that computes $f(x)$ in not more than $p(|x|)$ steps. Now $|x|$ is a lot smaller than x (with the exception of $x = 0$), and we can assume that the polynomial p has positive coefficients, and hence is monotonic. So the number of steps is bounded by $p(x)$, a primitive recursive function of x.

(Actually, we can get sharper bounds. For some constant c, we can get $p(|x|) \leq x + c$. But for our present purposes, the weaker result suffices.)   ⊣

But this is a very high upper bound; the converse to this result does not hold. We will see that many primitive recursive functions are *not* P-time computable. There is a limit to the growth rate of P-time computable functions, imposed by the fact that printing an output symbol takes a step. That is, we have the following constraint:

Growth limitation property: *If f is computable in time bounded by the polynomial p, then $|f(x)| \leq p(|x|)$.*

Proof. Initially, register 0 contains the empty word. Adding a symbol to register 0 requires at least one step. And we need to add $|f(x)|$ symbols.   ⊣

This prevents exponential functions from being P-time computable; there is not enough time to write down the result.

Lemma: *Where $|x|$ is the length of the binary numeral for x, the following hold for all $x \neq 0$.*

(a) $2^{|x|-1} \leq x \leq 2^{|x|} - 1$.
(b) $|x| = \lfloor \log_2 x \rfloor + 1$.

Proof.

(a) We must have x somewhere between the smallest $|x|$-bit number and the largest:

$$1\underbrace{00\cdots0}_{|x|-1} \leq x \leq 1\underbrace{11\cdots1}_{|x|-1}$$

And this is the inequality stated by (a).

(b) We can write part (a) as $2^{|x|-1} \leq x < 2^{|x|}$. To this inequality, apply the (monotonic) $\log_2$ function:

$$|x| - 1 \leq \log_2 x < |x|$$

Now round down: $|x| - 1 = \lfloor \log_2 x \rfloor$.

$\dashv$

Corollary: *The exponential function* $x \mapsto 2^x$ *is not P-time computable.*

Proof. How long would it take to write out 2^x? We have $|2^x| = x + 1$, and by the lemma this exceeds $2^{|x|-1}$. The function $t \mapsto 2^{t-1}$ grows faster than any polynomial. We conclude that the number of steps needed to write out 2^x cannot be bounded by a polynomial in $|x|$. (For example, if x is a 60-bit number, then 2^x has more than 2^{59} bits. At a million bits per second, it will take hundreds of centuries to write out 2^x.) $\dashv$

To balance this negative result, let's look at some functions that *can* be computed in P-time.

Example: The squaring function $x \mapsto x^2$ is P-time computable. First, think about how a *human* goes about squaring an n-bit number x. The standard procedure we all learned in the third grade involves building up an $n \times n$ array and then adding up the $2n$ columns. Building up the array is not difficult (multiplying by 0 or by 1 is easy), and the time it takes us will be proportional to n^2. Then adding up the $2n$ columns, each of height not more than n, will again take us some amount of time proportional to n^2. Altogether, the number of steps will be bounded by a quadratic in n.

A register machine is not a human, so there is more work to be done. To verify that squaring is P-time computable, we need to program the foregoing human procedure, and then obtain a bound on how long the program takes. The program can start by making a second copy of x (in $k|x|$ steps) and then calling the multiplication program outlined on page 77. The total number of steps will be bounded by a quadratic in $|x|$.

One way *not* to do squaring is by repeated addition. That is, a program that computes x^2 by adding x to itself x times will require exponential time (because x is about $2^{|x|}$) and $2^{|x|}$ grows faster than any polynomial in $|x|$.

Proposition: *The composition* $g \circ f$ *of P-time computable functions is again P-time computable.*

Proof. Assume that program $\mathcal{M}$ computes f in time bounded by the polynomial p, and that program $\mathcal{N}$ computes g in time bounded by the polynomial q. Then the program $\mathcal{M}$ followed by some minor housekeeping followed by $\mathcal{N}$ will compute $g \circ f$. How long does it take? The number of steps on input x is bounded by

$$p(|x|) + \text{small amount} + q(|f(x)|).$$

By the growth limitation property, $|f(x)| \leq p(|x|)$. We may assume that the polynomial q has only positive coefficients, and hence is monotonic. So our time bound becomes

$$p(|x|) + \text{small amount} + q(p(|x|))$$

which is a polynomial in $|x|$.   ⊣

For a function $f(x, y)$ of two variables to be considered P-time computable, there must be a polynomial p such that some program produces $f(x, y)$ in not more than $p(|x| + |y|)$ steps. That is, we can look at $|x| + |y|$ as the total length of the input. We can handle functions of more variables in a similar way.

The preceding proposition, regarding composition $g \circ f$, can be extended to cover compositions $g(f_1(\vec{x}), \dots, f_k(\vec{x}))$ of functions of several variables.

Example: The addition function $\langle x, y \rangle \mapsto x + y$ is P-time computable. As outlined on page 77, there is a program that will give $x + y$ in a number of steps bounded by $k \max(|x|, |y|)$ for a constant k. And $\max(|x|, |y|) \leq |x| + |y|$.

Example: The multiplication function $\langle x, y \rangle \mapsto xy$ is P-time computable. This is a lot like the squaring function.

Proposition: *Any polynomial function $p(x)$ is P-time computable.*

Proof outline. We build up the polynomial piece by piece, using the foregoing examples. Raising x to a power (that is, the function $x \mapsto x^k$) can be done by composition of multiplications. A monomial $x \mapsto cx^k$ involves one more multiplication, this time by a constant. Finally, we use a composition of additions.   ⊣

Example: Let

$$f(x) = 2^{|x|^2} = 100 \cdots 0_{\text{two}},$$

where the string of 0's is $|x|^2$ long. This function can be computed in polynomial time. To append a string of 0's of length $|x|$ to the output, we use a loop where each time through the loop we append one 0 and we erase one symbol from a copy of x. Now we put that loop inside another similar loop. Together, the loops append a string of 0's of length $|x|^2$, and they do this in quadratic (in $|x|$) time.

How fast does f grow? With help from a recent lemma, we obtain

$$f(x) = \left(2^{|x|}\right)^{|x|} \geq x^{|x|} \geq x^{\log_2 x}$$

so f grows faster than x^k for any fixed k. Thus, we have here a P-time computable function that grows faster than any polynomial p, in the sense that $f(x)/p(x) \to \infty$ as x increases.

We can conclude that the class of P-time computable functions is more than the class of polynomials, but less than the class of primitive recursive functions.

Often P-time computability is presented in terms of acceptance of languages (i.e., sets of words). We have a finite alphabet $\Sigma = \{0, 1\}$. Over this alphabet, the set Σ^*

of all words is the set of binary strings. (A binary string is the same thing as a base-2 numeral, except for the annoying problem of leading zeros. Ignoring that annoyance, we can think of Σ^* as being the same as the set of numerals for $\mathbb{N}$.) By a *language L*, we mean a set of words (i.e., a subset of Σ^*). We say that $L \in P$ if there is a program and a polynomial p such that the following hold:

- Whenever a word w is in L, then the program halts on input w (that is, it "accepts" w), and does so in not more than $p(|w|)$ steps.
- Whenever a word w is not in L, then the program never halts on input w (that is, the program does not accept w).

This definition is equivalent to one formulated in terms of P-time computable functions:

Theorem: *A language L is in P if and only if its characteristic function C_L is P-time computable.*

Proof idea. In one direction, this is easy: If C_L is P-time computable, then we can make an acceptance procedure that, given an input word w, computes $C_L(w)$ and then either halts or goes into an infinite loop.

It is the other direction that is interesting. Assume we have an acceptance procedure (which, in effect, is computing the semicharacteristic function c_L) that runs in time bounded by a polynomial p. We can add to the program an alarm clock that rings after time $p(|w|)$.

That is, given the input word w, we first compute $p(|w|)$. This does not take long; it can be done in a number of steps that is a polynomial in $||w||$. We store this number in a "timer" register.

Then we proceed as follows: With our right hand, we run the acceptance procedure on the word w; with our left hand, we decrement the timer. Or more precisely (because register machines don't have hands), we interleave two programs. In odd-numbered steps, we do the acceptance procedure; in even-numbered steps, we decrement the timer.

If and when the acceptance procedure halts (the timer will not have run out), then we give output Yes. If and when the timer runs out (the acceptance procedure will not have halted), we give output No. ⊣

As a corollary, we can conclude that whenever $L \in P$, then L, viewed as a set of numbers, is primitive recursive.

Of course, if the characteristic function of L is P-time computable, then so is the characteristic function of its complement, $\overline{L}$. So by the above theorem, $L \in P$ iff $\overline{L} \in P$. That is, $P = co\text{-}P$, where co-P is the collection of complements of languages in P.

Example: It is now known that the set of prime numbers, as a set of words written in the usual base-2 notation (or base-10, for that matter), belongs to P. Because the set of primes is in P, it follows that the set of composite numbers is in P as well.

Nonexample: Consider the function

$f(x) = $ the least prime divisor of x

with the convention that $f(0) = 0$ and $f(1) = 1$. It is easy to see that f is primitive recursive. Despite the fact that the set of primes belongs to P, it is currently an open question whether or not f is P-time computable.

The concept of belonging to P extends in a natural way to binary relations R on the set of words (i.e., $R \subseteq \Sigma^* \times \Sigma^*$).

Informally, L is in P if L is not only a decidable set of words, but moreover there is a "fast" decision procedure for L–one that we can actually implement in a practical way. For example, finite graphs can be coded by binary strings. The set of two-colorable graphs (i.e., the set of graphs that can be properly colored with two colors) is in P because coloring a graph with two colors does not involve any backtracking; either the coloring succeeds or we find a cycle of odd length. The set of graphs with an Euler cycle is in P because it is fast to check that the graph is connected and that every vertex has even degree.

What about three-colorable graphs, or graphs with Hamiltonian cycles? Here there are no known fast decision procedures. But there are weaker facts: Given a proper coloring with three colors, it is fast to verify that it is indeed a proper coloring. Given a Hamiltonian cycle, it is fast to verify that it is indeed Hamiltonian. Both the set of three-colorable graphs and the set of Hamiltonian graphs are examples of languages that are "verifiable" in P-time. That is, we might not know fast decision procedures, but we do know how, given the correct evidence, to verify quickly that the evidence does indeed show that the graph has the claimed property. Such languages belong to a class known as NP.

One way to define NP is to use *nondeterministic* Turing machines. (The symbols "NP" stand for "nondeterministic polynomial time.") Back in Chapter 1, the definition of a Turing machine demanded that a machine's table of quintuples be unambiguous, that is, that no two different quintuples have the same first two components. By simply omitting that demand, we obtain the concept of a nondeterministic Turing machine. A computation of such a machine $\mathcal{M}$, at each step, is allowed to execute *any* quintuple that begins with its present state and the symbol being scanned. So when we start $\mathcal{M}$ on some input, there can be many possible computations, depending on which of the allowed quintuples it chooses to execute.

Then we say that $L \in$ NP if there is a nondeterministic Turing machine M and a polynomial p, such that the following conditions hold:

- Whenever a word w is in L, then *some* computation of $\mathcal{M}$ starting from input w halts, and does so in not more than $p(|w|)$ steps.
- Whenever a word w is not in L, then *no* computation of $\mathcal{M}$ starting from input w *ever* halts.

An accepting computation can be thought of as having made a number of lucky guesses.

There is an equivalent, and somewhat more workable, characterization along the lines of Σ_1 definability.

Definition: For a language L, $L \in$ NP if there is binary relation $R \in$ P and a polynomial p, such that for every word w,

$$w \in L \iff \exists y[|y| \leq p(|w|) \text{ and } R(w, y)].$$

Example: The set of three-colorable graphs (as a set of binary strings) is in NP. A graph w is three-colorable iff there exists some three-coloring y, such that w is properly colored by y (that is, adjacent vertices are always different colors).

Similarly, the set of Hamiltonian graphs is in NP. Here, the evidence y is a Hamiltonian cycle in the graph w.

Another example of a language in NP is SAT, the set of satisfiable formulas of sentential logic. The truth-table method taught in logic courses for determining whether a formula with n sentence symbols is satisfiable involves forming all 2^n lines of the formula's truth table, and looking to see if there is a line making the formula true. But this is not a feasible algorithm because 2^{80} microseconds greatly exceeds the age of the universe. But if we (nondeterministically) guess the correct line of the table, then we can quickly verify that the formula is true under that line.

There is a clear analogy between computable and recursively enumerable (r.e.) sets on the one hand, and P and NP on the other hand. The computable sets are decidable; the sets in P are decidable by fast algorithms. And r.e. sets are one existential quantifier away from being computable; sets in NP are one existential quantifier away from being in P. Moreover, there are r.e. sets that are complete with respect to $\leq_m$; there are NP sets with a similar property. Say that L_1 is P-time *reducible* to L_2 (written $L_1 \leq_P L_2$) if there is a P-time computable (total) function f that many-one reduces L_1 to L_2. The following result was proved independently by Cook (1971) and Leonid Levin (1973):

Cook–Levin theorem: *SAT is in NP, and every NP language is P-time reducible to SAT.*

In other words, SAT is NP-complete. Karp has shown that many other NP languages (three-colorable graphs, Hamiltonian graphs, and others) are NP-complete.

Digression: We have defined P-time reducibility in a way that is analogous to many-one reducibility. But there is another option: We could define a concept analogous to Turing reducibility. That is, we could specify that L_1 be polynomial-time decidable, but allow an oracle for L_2.

7.2 P versus NP

How far does the analogy between "NP" and "r.e." go? We know that there are noncomputable r.e. sets, and a set is computable if and only if both it and its complement are r.e. Although it is clear that $P \subseteq NP \cap \text{co-NP}$ (that is, every language in P is also in NP, as is its complement), it is not known whether $P = NP$, or if NP is closed under complements.

The diagonalization that produces a noncomputable r.e. set K was "relativized" in Chapter 6 to show that for any fixed oracle B, there is a set B' that is r.e. relative to B but is not computable relative to B. Might some diagonal argument produce a set in NP that was not in P? Would that argument then relativize? The definitions of P and NP extend easily to P^B and NP^B, where the computations can query the oracle B (in one step).

In a 1975 article, Theodore Baker, John Gill, and Robert Solovay showed that there are oracles B and C such that on the one hand $P^B = NP^B$ and on the other hand $P^C \neq NP^C$. This result suggests that the "P versus NP" question is difficult because whatever argument might settle the question cannot relativize in a straightforward way. It has also been shown that if we choose the oracle B at *random* (with respect to the natural probability measure on $\mathcal{P}\mathbb{N}$), then $P^B \neq NP^B$ with probability 1.

The P versus NP question remains the outstanding problem in theoretical computer science. In recognition of this fact, the Clay Mathematics Institute has offered a million-dollar prize for its solution.

7.3   Some Other Complexity Classes

And what might lie beyond NP? Although there is some analogy between NP and Σ_1, what might be analogous to Σ_n? And what might be computable in "exponential time," where we allow the computing time to be bounded by $2^{p(|x|)}$ for a polynomial p?

As indicated earlier, two reasonable measurements of complexity are *time* (the number of steps the computation executes before halting) and *space* (where we take the largest number of symbols a register ever contains in the course of the computation, and add these numbers up for all the registers).

These two measurements are related. If a computation halts in time t on input x, then the space used is bounded by $t + |x|$ because writing a symbol takes a step. What about the other direction? Suppose that a particular calculation from a program halts, having used space s. We want a bound on the time it used.

Consider the snapshots

[location counter, memory number]

that arise in the course of the computation. One thing we can be sure of is that no snapshot occurs twice. This is because if a computation ever hits a snapshot for a second time, then the computation will run forever, returning to this snapshot infinitely often. So for a calculation that halts, the time is bounded by the number of possible snapshots. And what is the number of possible snapshots? For a program with Gödel number e, there are at most $1 + \text{lh}\, e$ values for the location counter. And if the program addresses k different registers, there could be (for the alphabet $\{0, 1\}$) at most 2^{ks} different values for the memory number. Putting the pieces together, we see that for constants c and k (depending on the program), the running time is bounded by $c2^{ks}$.

We have been looking at the complexity class P. This class might just as well be called PTIME because it uses time as the complexity measure. A language L (that is, a subset of $\{0, 1\}^*$) is in the class if its characteristic function at x is computable in *time* bounded by a polynomial $q(|x|)$ in $|x|$.

Analogously, define PSPACE as follows: A language L belongs to PSPACE if there is a program and a polynomial p such that the program computes the characteristic function of L at each word x using *space* at most $p(|x|)$. Then $P \subseteq PSPACE$ because a computation that uses time $q(|x|)$ can use at most space $|x| + q(|x|)$.

Next, we want to define EXPTIME and EXPSPACE. A language L belongs to EXPTIME if there is a program and a polynomial p such that the program computes the characteristic function of L at each word x in at most $2^{p(|x|)}$ steps. And a language L belongs to EXPSPACE if there is a program and a polynomial p, such that the program computes the characteristic function of L at each word x using *space* at most $2^{p(|x|)}$.

We claim that PSPACE $\subseteq$ EXPTIME. Suppose a computation uses space $p(|x|)$. Then for some constants, c and k, that depend on the program (but not on x), the running time is bounded by $c2^{kp(|x|)} = 2^{\ln c + kp(|x|)}$. We observe that the exponent here is a polynomial in $|x|$.

Moreover, EXPTIME $\subseteq$ EXPSPACE. Suppose a computation uses time $2^{p(|x|)}$. Then the space is bounded by $|x| + 2^{p(|x|)}$. This is in turn bounded by $2^{q(|x|)}$ for a suitable polynomial q; see Exercise 4.

Thus, we are left with the inclusions

$$P \subseteq PSPACE \subseteq EXPTIME \subseteq EXPSPACE \subseteq PR,$$

where PR is the class of primitive recursive languages. There are numerous questions that can be raised regarding these classes. For some of the questions, answers are known, whereas other questions remain open.

The subject of computational complexity is young and growing. See the list of references for some avenues well worth exploring.

Exercises

1. Show that the concept of P-time reducibility is reflexive and transitive. That is, show that we always have $L \leq_P L$, and that whenever $L_1 \leq_P L_2$ and $L_2 \leq_P L_3$, then $L_1 \leq_P L_3$.
2. Assume that $L_1 \leq_P L_2$ and $L_2 \in P$. Show that $L_1 \in P$.
3. Assume that $L_1 \leq_P L_2$ and $L_2 \in NP$. Show that $L_1 \in NP$.
4. Assume that p is a polynomial. Show that for some polynomial q, we have $z + 2^{p(z)} \leq 2^{q(z)}$ for all z.

A1 Mathspeak

The purpose of this appendix is to give a quick summary of everyday mathematical terminology.

First of all, a *set* is a collection of things (called its *members* or *elements*), the collection being regarded as a single object. We write "$x \in S$" to say that x is a member of S; we write "$x \notin S$" to say that it is not.

For example, there is the set S whose members are the prime numbers between 0 and 10. This set has four elements, the numbers 2, 3, 5, and 7. We can name this set conveniently by listing the members within braces (curly brackets):

$$S = \{2, 3, 5, 7\}$$

An important part of the set concept is that what a set *is* depends solely on what its members are, and *not* on how we might choose to name them. We might choose to name the elements in a different order

$$\{2, 3, 5, 7\} = \{7, 5, 3, 2\}$$

or even with repetitions

$$\{2, 3, 5, 7\} = \{2, 3, 3, 5, 5, 5, 7, 7, 7, 7\}.$$

In each case, we are referring to one and the same set. Something belongs to the set on the left iff it belongs to the set on the right (where "iff" abbreviates "if and only if"). We might even describe S as the set of all solutions to the polynomial equation

$$x^4 - 17x^3 + 101x^2 - 247x + 210 = 0.$$

No matter; it is still the same set.

Similarly, we can name larger sets. $\{0, 2, 4, \ldots, 20\}$ is the set of even natural numbers up to 20, and

$$\mathbb{N} = \{0, 1, 2, \ldots\}$$

is the infinite set of all natural numbers. Similarly, the set $\mathbb{Z}$ of all integers can be expressed by

$$\mathbb{Z} = \{\ldots -2, -1, 0, 1, 2, \ldots\}.$$

And we can name smaller sets; {2} is the one-element set (the *singleton*) whose only member is 2. (This set is not the same thing as the number 2. A hatbox containing one hat is not to be worn on the head.)

There is an even smaller set, the zero-element set with no members at all. It is conventional to call this set (and there can be only one such set) ∅.

Another convenient way to name the set of all objects that meet some condition _____ is to write

$$\{x \mid \underline{\quad x \quad}\}$$

as in

$$\{2, 3, 5, 7\} = \{x \mid x^4 - 17x^3 + 101x^2 - 247x + 210 = 0\}.$$

Here "x" is a dummy variable; we could just as well write

$$\{2, 3, 5, 7\} = \{t \mid t^4 - 17t^3 + 101t^2 - 247t + 210 = 0\}.$$

This notation is convenient for sets that are or might be infinite, as in

$$\{x \mid x \text{ and } x + 2 \text{ are both prime}\}.$$

Or to clarify that we mean here a set of natural numbers, we can write

$$\{x \in \mathbb{N} \mid x \text{ and } x + 2 \text{ are both prime}\}.$$

For example, we can use this notation to describe intersections

$$A \cap B = \{x \mid x \in A \text{ and } x \in B\}$$

and unions

$$A \cup B = \{y \mid y \in A \text{ or } y \in B \text{ (or both)}\}.$$

Similarly, we can describe the "relative complement"

$$A \setminus B = \{t \mid t \in A \text{ and } t \notin B\}.$$

We say that a set S is a *subset* of a set T or that T *includes* S (written $S \subseteq T$) if all members of S (if any) are also members of T. For any set S, we have both $S \subseteq S$ and $\emptyset \subseteq S$. The latter is "vacuously true," in the sense that the task of verifying, for each member of $\emptyset$, that it also belongs to S, requires doing nothing at all. A *proper subset* of S is a subset of S different from S itself.

The subset relation ($\subseteq$) is not to be confused with the membership relation ($\in$). If we want to know whether $S \in T$, we look at S as a single object, and we check to see if this single object is among the members of T. By contrast, if we want to know whether $S \subseteq T$, then we need to open up the set S, examine its various members (if any), and check to see if each of them is also among the members of T.

For a set S, we can form a new set, called the power set of S, whose members are exactly the subsets of S. So the power set, call it $\mathcal{P}S$, is always a set of sets. For example, if S is the three-element set $S = \{1, 2, 3\}$, then its power set has size eight:

$$\mathcal{P}S = \{\emptyset, \{1\}, \{2\}, \{3\}, \{1, 2\}, \{1, 3\}, \{2, 3\}, \{1, 2, 3\}\}.$$

As sets, $\{2, 3\} = \{3, 2\}$; we have at least these two names for the set. But sometimes, we *want* to keep track of the order. That is, we will want the *ordered pair* $\langle x, y \rangle$, consisting of x and y and a designation of x as first and y as second. Then the four ordered pairs

$$\langle 2, 3 \rangle, \quad \langle 3, 2 \rangle, \quad \langle 2, 2 \rangle, \quad \langle 3, 3 \rangle$$

are all different.

For sets A and B (not necessarily different), the set of *all* ordered pairs $\langle x, y \rangle$ with x from A and y from B is called the Cartesian product $A \times B$ of A and B:

$$A \times B = \{\langle x, y \rangle \mid x \in A \text{ and } y \in B\}$$

Similarly, we can use ordered triples $\langle x, y, z \rangle$, ordered quadruples $\langle u, v, x, y \rangle$, ordered quintuples $\langle u, v, x, y, z \rangle$, and so forth for "tuples" of other lengths. In general, $\langle x_1, x_2, \ldots, x_n \rangle$ is an ordered n-tuple. For a set S, we let S^n be the set of all ordered n-tuples $\langle s_1, s_2, \ldots, s_n \rangle$, where each s_i is in S. (There is some rationale to this notation. If S is a set of size 17, then S^3 is a set of size $17^3 = 4913$.)

A *relation* is defined to be a set of ordered pairs. That is, a set R is a relation if each of its members (if any) is an ordered pair. This usage of the word "relation" has some connection with the everyday usage of the word. For example, the ordering relation on the set $\mathbb{R}$ of real numbers is completely described by the set of ordered pairs $\langle x, y \rangle$ of reals with $x < y$, that is,

$$\{\langle x, y \rangle \in \mathbb{R}^2 \mid x < y\}.$$

It is now a small additional step to say that this set of pairs *is* the ordering relation. R is a relation *on* S if it is a set of ordered pairs of members of S, that is, if $R \subseteq S^2$.

We can stretch this terminology further: A *ternary relation* is a set of ordered triples. An n-ary relation on a set S is a subset of S^n. For clarity, a set of ordered pairs can be called a *binary* relation.

In Chapter 6, we encounter *equivalence relations* on a set. These are relations with certain special properties.

A *function* f on a set S assigns to each member x of S one and only object, which is written $f(x)$. Functions are familiar from calculus courses. For example, calculus gives us the chain rule for the derivative of the *composition* of functions. The set S is called the *domain* of the function f. For some object x *not* in the domain of f, one can say that $f(x)$ is "undefined." For example, calculus defines the logarithm function as a function on the set of positive real numbers. But $\log(-3)$ is undefined.

The *graph* of a function f is the set of all ordered pairs $\langle x, y \rangle$ for which $y = f(x)$. Thus for a function f on S, the graph is the set

$$\{\langle x, f(x) \rangle \mid x \in S\}$$

of pairs. So it is a relation. Often people take a function simply to *be* this relation. Doing so has the advantage that the domain of the function f is nothing but the set of objects that are first in an ordered pair in f. And the *range* is nothing but the set of objects that are second. It is convenient to write $f : A \to B$ (and to say that f *maps A into B*) to mean that f is a function with domain A and with range included in B:

$$f : A \to B \iff f \text{ is a function and } \operatorname{dom} f = A \text{ and } \operatorname{ran} f \subseteq B$$

If, in addition, we have $\operatorname{ran} f = B$, then we say that f maps A *onto* B.

A function f is said to be *one-to-one* if whenever we take two different objects, say x and y, in the domain of f, then the values $f(x)$ and $f(y)$ are different. For example, in calculus, the exponential function $f(x) = e^x$ is one-to-one, but the sine function is not. For a one-to-one function f, we can construct its *inverse f^{-1}*:

$$f^{-1} : \operatorname{ran} f \to \operatorname{dom} f,$$

where $f^{-1}(y)$ is the unique x in $\operatorname{dom} f$ for which $f(x) = y$.

Digression: If one were inventing mathematical notation from scratch, one should write "$(x)f$" instead of "$f(x)$." That way, certain equations related to the composition of functions would read more smoothly (in languages in which one reads from left to right). Too late now.

On a similar subject, a flaw in standard calculus notation is that it lacks a good symbol for the identity function. Instead, people are forced into circumlocutions such as "the function whose value at t is t," where t is a dummy variable. The squaring function is the function whose value at x is x^2. How awkward! The notation $x \mapsto x^2$ helps a little.

A2 Countability

Often we want to know the *size* of a set. On the one hand, there are the finite sets. On the other hand, there are the infinite sets. The infinite sets are bigger than the finite sets.

There is more to it, of course. There is the zero-element set, the empty set. There are the one-element sets, the singletons (like $\{8\}$). There are the two-element sets, the doubletons (like $\{0, 8\}$). And so forth and so on. Finite sets come in all sizes.

Something similar happens with the infinite sets. All the infinite sets are big, but some are bigger than others. We want to make sense of this idea, by extending some concepts (that are familiar in the finite case) to infinite sets.

For sets A and B, say that A is the *same size* as B (written $A \approx B$) if there is a one-to-one correspondence between them, that is, if there is a one-to-one function f whose domain is A and whose range is B. (In this situation, f^{-1} is a one-to-one function, whose domain is B and range is A. Hence B is also the same size as A, so the concept is symmetric.)

Applied to finite sets, this concept tells us nothing much that is new. For infinite sets, the situation is more interesting. The possibly surprising fact about infinite sets is that they are *not* all the same size.

One infinite set is the set $\mathbb{N} = \{0, 1, 2, \ldots\}$ of all natural numbers. We can use natural numbers to give an exact characterization of finiteness: A set is finite iff there is a natural number n such that the set is the same size as $\{x \in \mathbb{N} \mid x < n\}$, that is, the same size as $\{0, 1, \ldots, n - 1\}$. (For the empty set, $n = 0$.)

Definition: A set is said to be *countable* if it is the same size as some subset of $\mathbb{N}$. That is, a set S is countable if there is a one-to-one function $f : S \to \mathbb{N}$ mapping S into the natural numbers, so that $S \approx \operatorname{ran} f$. Otherwise, the set is said to be *uncountable*.

Thus, for a set S to be countable, there must be a way to assign a unique natural number to each member of S. For example, any finite set is countable because it has the same size as $\{0, 1, \ldots, n - 1\}$, for some n. And $\mathbb{N}$ itself is countable, as are each of its subsets.

Example: The set of all finite sequences of natural numbers is countable. In Chapter 2, we defined the bracket notation:

$$[\,] = 1$$
$$[x] = 2^{x+1}$$
$$[x, y] = 2^{x+1}3^{y+1}$$
$$[x, y, z] = 2^{x+1}3^{y+1}5^{z+1}$$
$$\cdots$$
$$[x_0, x_1, \ldots, x_k] = 2^{x_0+1}3^{x_1+1} \cdots p_k^{x_k+1}$$

The function

$$\langle x_0, x_1, \ldots, x_k \rangle \mapsto [x_0, x_1, \ldots, x_k]$$

maps the set of sequences of natural numbers into $\mathbb{N}$, and it is one-to-one by the uniqueness of prime factorization.

Theorem: *Any infinite countable set has the same size as* $\mathbb{N}$.

Thus, the countable sets consist of the finite sets,

$$\{s_0, s_1, \ldots, s_{n-1}\}$$

plus the sets

$$\{s_0, s_1, \ldots, \}$$

that are the same size as $\mathbb{N}$.

Proof. Assume that S is an infinite set that is countable, so that there is a one-to-one function $f : S \to \mathbb{N}$. We want a new function $g : S \to \mathbb{N}$ that is both one-to-one and *onto* $\mathbb{N}$. That is, we know that ran $f \subseteq \mathbb{N}$, and we want ran $g = \mathbb{N}$. The idea is to push down ran f, to squeeze out all the holes.

First of all, ran f contains some least member, say $f(s_0)$. (Because f is one-to-one, s_0 is unique.) We define $g(s_0) = 0$. More generally, for each n, there is a unique s_n in S for which $f(s_n)$ is the $(n + 1)$st member of ran f. We define $g(s_n) = n$. This gives us the function g we want: dom $g = S$ and ran $g = \mathbb{N}$.   ⊣

Theorem:
(a) *Any subset of a countable set is countable.*
(b) *The union of two countable sets is countable.*
(c) *The Cartesian product of two countable sets is countable.*
(d) *If A is a countable set, then the set A^* of all finite sequences of members of A is countable.*
(e) *The union of countably many countable sets is countable.*

Proof. The preceding example proves part (d) in the special case, where $A = \mathbb{N}$. The argument can be adapted to cover any countable A.

As a special case of part (c), the set $\mathbb{N} \times \mathbb{N}$ is countable; we can map the ordered pair $\langle x, y \rangle$ to $2^{x+1}3^{y+1}$ as before. (And there are other possible "pairing functions," as noted on page 43. For a start, we could use $2^x 3^y$, which is a bit simpler.)

Again, the argument can be adapted to cover the Cartesian product $A \times B$ of any countable sets A and B. Where f and g are one-to-one functions with $f : A \to \mathbb{N}$ and $g : B \to \mathbb{N}$, we can map the ordered pair $\langle a, b \rangle$ to $2^{f(a)}3^{g(b)}$.

Moreover, in this situation, the union $A \cup B$ is countable. We can map x to $2f(x)$ whenever $x \in A$, and to $2g(x) + 1$ when $x \notin A$. Thus, we get part (b).

Part (a) of the theorem is easy to see.

Part (e) means the following: Assume that $\mathcal{A}$ is countable, and that each member of $\mathcal{A}$ is a countable set (so in particular, $\mathcal{A}$ is a set of sets). Then part (e) says that $\bigcup \mathcal{A}$, the result of dumping all the members of $\mathcal{A}$ together, is countable.

We may suppose that $\mathcal{A}$ is infinite (otherwise, we can simply apply part (b) several times) so there is some function $f : \mathcal{A} \to \mathbb{N}$ that is both one-to-one and onto $\mathbb{N}$:

$$\mathcal{A} = \{f(0), f(1), \ldots, f(n), \ldots\}.$$

For each n, the set $f(n)$ is countable, so there exists some function mapping it one-to-one into $\mathbb{N}$. We need to *choose* some such function g_n for each n. Then for each x in $\bigcup \mathcal{A}$, we take the smallest n for which $x \in f(n)$ and map x to the natural number $2^n 3^{g_n(x)}$. The map described in this way maps $\bigcup \mathcal{A}$ one-to-one into $\mathbb{N}$.   ⊣

For example, the set $\mathbb{Z}$ of all integers (positive, negative, and zero) is a countable set. And the set $\mathbb{Q}$ of all rational numbers is countable. Part (d) tells us that over a countable alphabet A, the set A^* of all words is countable.

But not every set is countable. And by part (a) of the theorem, any set having an uncountable subset must be uncountable.

Cantor's theorem (1873):

(a) *The set $\mathbb{R}$ of all real numbers is uncountable.*
(b) *The set $\mathcal{P}\mathbb{N}$ of all subsets of $\mathbb{N}$ is uncountable.*
(c) *The set of all infinite binary sequences (i.e., the set of all functions from $\mathbb{N}$ into $\{0, 1\}$) is uncountable.*
(d) *The set $\mathbb{N}^{\mathbb{N}}$ of all function from $\mathbb{N}$ into $\mathbb{N}$ is uncountable.*

Proof. This theorem is proved by the classical "Cantor diagonal argument." To show that a set is uncountable, it suffices to show that each countable subset fails to exhaust the set.

For part (a), consider an arbitrary countable set of real numbers, for example:

$$s_0 = 236.001 \cdots$$
$$s_1 = -7.777 \cdots$$
$$s_2 = 3.1415 \cdots$$

To show that this list fails to exhaust $\mathbb{R}$, we only need to produce one new real number z not on the list. Here is one: Its integer part is 0, and for each n, its $(n + 1)$st decimal place is 7 unless the $(n + 1)$st decimal place of s_n is 7, in which case the $(n + 1)$st decimal place of z is 5. So in the example shown, $z = 0.757 \cdots$. Then z cannot have been on the list because it differs from each s_n in its $(n + 1)$st decimal place.

To prove part (b), consider an arbitrary countable subset

$$\{S_0, S_1, \ldots\}$$

of $\mathcal{P}\mathbb{N}$. To show that this collection does not exhaust $\mathcal{P}\mathbb{N}$, we seek to come up with a new subset of $\mathbb{N}$. Here is one: $A = \{n \in \mathbb{N} \mid n \notin S_n\}$. This set could not equal S_{17}

because either

$$17 \in S_{17} \text{ and } 17 \notin A \qquad \text{or} \qquad 17 \notin S_{17} \text{ and } 17 \in A.$$

The set in part (c) has the same size as the set in part (b); simply pair up each subset of $\mathbb{N}$ with its characteristic function. Or to prove part (c) directly, consider any countable set $\{s_0, s_1, \ldots\}$ of infinite binary sequences. Then, we can make a new binary sequence f by defining $f(n) = 1 \doteq s_n(n)$ for each n.

The set in part (d) is at least as big as the set in part (c). That is, the set in part (c) is a subset of the set in part (d).   ⊣

A particularly relevant example for our purposes is the set $\mathcal{S}$ of all register-machine programs. This set is countable. One way to see this fact is to represent $\mathcal{S}$ as a set of finite sequences over a certain finite alphabet. But a more direct proof uses the function $\mathcal{P} \mapsto \#\mathcal{P}$ assigning to each program its Gödel number. This function maps $\mathcal{S}$ one-to-one into $\mathbb{N}$.

Consequently, the set of all computable partial functions is countable. We can map each such function to the least Gödel number of a program that computes it.

The set of *all* partial functions (computable or not) is uncountable. By part (d) of the preceding theorem, even the set of total functions is uncountable. So the set of *noncomputable* total functions is uncountable. That is, there are uncountably many noncomputable functions.

A3 Decadic Notation

There is a simple and natural one-to-one correspondence between the set of all strings over a finite alphabet and the set of natural numbers. The key to the correspondence is to use base-n notation, where n is the size of the alphabet, but without a 0 digit.

Suppose that Σ is a finite set of size n. We refer to Σ as the *alphabet*, we refer to the members of Σ as *letters*, and we refer to finite strings of letters as *words* (over Σ). Let Σ^* be the infinite set of all words (including the empty word λ). For example, if $\Sigma = \{1, 2\}$, then

$$\Sigma^* = \{\lambda, 1, 2, 11, 12, 21, 22, 111, 112, 121, \ldots\}.$$

We assume that the members of the alphabet Σ are ordered in some way (referred to as *alphabetic order*). Thus, we have a function $v : \Sigma \to \{1, \ldots, n\}$ where

$$v(\text{the first letter}) = 1$$
$$v(\text{the second letter}) = 2$$
$$\cdots$$
$$v(\text{the last letter}) = n,$$

and we refer to $v(a)$ as the *value* of the letter a.

We obtain a one-to-one map from Σ^* onto the set $\mathbb{N} = \{0, 1, 2, \ldots\}$ by mapping the three-letter word abc to the number

$$(abc)_{n\text{-adic}} = v(a)n^2 + v(b)n + v(c)$$

and in general,

$$(a_k \cdots a_2 a_1)_{n\text{-adic}} = v(a_k)n^{k-1} + \cdots + v(a_2)n + v(a_1)$$

and $(\lambda)_{n\text{-adic}} = 0$ (the empty sum).

In what follows, the properties of this map will be developed in the special case where $n = 10$ and the 10 letters are $1, 2, 3, 4, 5, 6, 7, 8, 9, X$, in that order. But all of the arguments generalize immediately to finite alphabets of any size 2 or more.

(An alphabet of size 1 is a somewhat special and somewhat boring case. Where the alphabet is the singleton $\{|\}$, a typical word is the string $|||||||$, and the above equation reduces to

$$(a_k \cdots a_2 a_1)_{1\text{-adic}} = 1 + \cdots + 1 + 1 = k$$

so that $(w)_{\text{1-adic}}$ equals the length of the word w. This obviously produces a one-to-one map onto $\mathbb{N}$.)

In the $n = 10$ case, we will refer to $(w)_{\text{10-adic}}$ as $(w)_{\text{decadic}}$, the number denoted by the word w in *decadic* notation. (The corresponding words for $n = 1, 2, 3, \ldots$ would be monadic, dyadic, triadic, ...) For example,

$$(\mathbf{415})_{\text{decadic}} = 4 \cdot 100 + 1 \cdot 10 + 5 = 415$$

and

$$(\mathbf{4X5})_{\text{decadic}} = 4 \cdot 100 + 10 \cdot 10 + 5 = 505.$$

As the first of these equations exemplifies, for any word w not containing the X digit, $(w)_{\text{decadic}}$ is simply the number denoted by the numeral w in the usual base-10 notation. But the alphabet contains no zero digit, and some words contain the X digit (the "ten" digit). For example,

$$(\mathbf{XX})_{\text{decadic}} = 10 \cdot 10 + 10 = 110$$

and

$$(\mathbf{XXX})_{\text{decadic}} = 10 \cdot 100 + 10 \cdot 10 + 10 = 1110.$$

First consider the problem of how to add 1 in decadic notation. Here are some examples:

$$(\mathbf{3X8})_{\text{decadic}} + 1 = (\mathbf{3X9})_{\text{decadic}}$$
$$(\mathbf{XXX})_{\text{decadic}} + 1 = (\mathbf{1111})_{\text{decadic}}$$
$$(\mathbf{2XXX})_{\text{decadic}} + 1 = (\mathbf{3111})_{\text{decadic}}$$

From looking at these examples, we want to extract a general rule for adding 1. In general, a word w consists of a word u (possibly empty) not ending in X, followed by a string of k X's (where $k \geq 0$). Thus,

$$(w)_{\text{decadic}} = (u)_{\text{decadic}} \cdot 10^k + 10^k + 10^{k-1} + \cdots + 10.$$

In order to add 1, we use the word w^+ consisting of u^+ followed by a string of k 1's, where u^+ is obtained by incrementing u's rightmost digit (and $\lambda^+ = 1$). This works because

$$(w^+)_{\text{decadic}} = ((u)_{\text{decadic}} + 1) \cdot 10^k + 10^{k-1} + \cdots + 10 + 1$$
$$= (u)_{\text{decadic}} \cdot 10^k + 10^k + 10^{k-1} + \cdots + 10 + 1$$
$$= (w)_{\text{decadic}} + 1.$$

Now that we know how to add 1, we obtain a theorem:

Theorem: *For every natural number n, we can find a word w for which* $(w)_{decadic} = n$.

Proof. We prove the existence of *w* by induction on *n*. The basis, *n* = 0, holds because $(\lambda)_{decadic} = 0$. For the inductive step, we add 1 as above.

And not only does *w* exist, but we know how to calculate it by adding 1 many times (*n* times). (Of course, there are much faster ways to find *w*, if we are in a hurry.)   ⊣

Next consider the problem of comparing decadic numerals to see which one denotes a larger number. Among four-digit numbers (i.e., among the numbers denoted by four-digit words), the smallest is obviously $(\mathbf{1111})_{decadic}$. Any other four-digit word will give us a larger number. Similarly, the largest number denoted by a three-digit word is $(\mathbf{XXX})_{decadic}$. Any other three-digit word will give us a smaller number. Moreover,

$$(\mathbf{XXX})_{decadic} < (\mathbf{1111})_{decadic}$$

because adding 1 to the left side gives the right side.

We can conclude from this that any number *m* denoted by a three-digit word is less than any number *n* denoted by a four-digit word:

$$m \le (\mathbf{XXX})_{decadic} < (\mathbf{1111})_{decadic} \le n$$

And the argument generalizes: Any number denoted by a *k*-digit word is less than any number denoted by a $(k + 1)$-digit word. And then by iterating this argument, we see that any number denoted by a *k*-digit word is less than any number denoted by a word of more than *k* digits. That is, shorter words always give smaller numbers.

What, then, about two words of the same length? For example, take the four-letter words **2∗∗∗** and **3★★★**, where in both cases, we don't know the last three digits. Even so, we know that

$$\begin{aligned} \mathbf{2*\!*\!*} &\le \mathbf{2XXX} \\ &< \mathbf{3111} \quad \text{by 1} \\ &\le \mathbf{3\!\star\!\star\!\star}. \end{aligned}$$

Similary, for the six-letter words **472∗∗∗** and **473★★★**, we have

$$\begin{aligned} \mathbf{472*\!*\!*} &\le \mathbf{472XXX} \\ &< \mathbf{473111} \quad \text{by 1} \\ &\le \mathbf{473\!\star\!\star\!\star}. \end{aligned}$$

In general, we need to look only at the first (i.e., leftmost) digit where two words of the same length disagree. The larger digit produces a larger number, no matter what the later digits are. That is, for words of the same length, we simply use lexicographic order. We can summarize these ideas as follows:

Theorem:

(a) *For two words of different lengths, the shorter word denotes a smaller number than does the longer word.*

(b) *For two words u and w of the same length, $(u)_{decadic} < (w)_{decadic}$ iff u lexicographically precedes w.*

In particular, two *different* words must denote different numbers, because either they will have different lengths (and clause (a) of the theorem will apply) or else there will be a first digit where they differ (and clause (b) will apply). That is, our map from words to numbers is one-to-one. Together with the previous theorem, we now have the following:

Theorem: *Decadic notation yields a one-to-one correspondence between the set of all words over our 10-letter alphabet and the set $\mathbb{N}$ of all natural numbers.*

This theorem illustrates a property of decadic notation which standard decimal notation lacks. In decimal notation, there is the problem of "leading zeros"; the words **3** and **03** denote the same number (or else **03** needs to be declared an illegal word).

Roman numerals are sometimes criticized for lacking a numeral for zero. But the real difficulty with Roman numerals is the lack of place-value notation. Zero itself is nothing.

References

1. Turing AM. On computable numbers, with an application to the Entscheidungsproblem. *Proc Lond Math Soc.* 1936–7;s2-42:230–265, 1937;43:544–546.
2. Rogers Jr H. *Theory of Recursive Functions and Effective Computability.* McGraw-Hill; 1967, MIT Press; 1987.
3. Church A. An unsolvable problem of elementary number theory. *Am J Math.* 1936;58: 345–363.
4. Shepherdson JC, Sturgis HE. Computability of recursive functions. *J Assoc Comput Mach.* 1963;10:217–255.
5. Post EL. Recursively enumerable sets of positive integers and their decision problems. *Bull Am Math Soc.* 1944;50:284–316.
6. Turing AM. Systems of logic based on ordinals. *Proc Lond Math Soc.* 1939;s2-45(3): 161–228.
7. Baker T, Gill J, Solovay R. Relativizations of the P = ? N P question. *SIAM J Comput.* 1975;4:431–442.
8. Agrawal M, Kayal N, Saxena N. PRIMES is in P. *Ann Math.* 2004;160:781–793.
9. Arora S, Barak B. *Computational Complexity: A Modern Approach.* Cambridge University Press; 2009.
10. Barwise J, ed. *Handbook of Mathematical Logic.* North-Holland Publishing Co.; 1978.
11. Barwise J, Etchemendy J. *Turing's World 3.0 for the Macintosh: An Introduction to Computability Theory.* CSLI; 1993.
12. Blum M. A machine-independent theory of the complexity of recursive functions. *J Assoc Comput Mach.* 1967;14:322–336.
13. Church A. A note on the Entscheidungsproblem. *J Symbolic Logic.* 1936;1:40–41, 101–102.
14. Cooper Barry S. *Computability Theory.* Chapman & Hall/CRC; 2003.
15. Cutland N. *Computability: An Introduction to Recursive Function Theory.* Cambridge University Press; 1980.
16. Davis M, ed. *The Undecidable: Basic Papers on Undecidable Propositions, Unsolvable Problems and Computable Functions.* Raven Press; 1965.
17. Davis M, Weyuker E. *Computability, Complexity, and Languages.* Academic Press; 1983. Second edition (with Ron Sigel), 1994.
18. Dietzfelbinger M. *Primality Testing in Polynomial Time—from Randomized Algorithms to "PRIMES is in P."* Lecture Notes in Computer Science, vol. 3000, Springer-Verlag; 2004.
19. Enderton HB. *A Mathematical Introduction to Logic.* Academic Press; 1972. Second edition, 2001.
20. Garey MR, Johnson DS. *Computers and Intractability: A Guide to the Theory of NP-Completeness.* W. H. Freeman; 1979.
21. Griffor E, ed. *Handbook of Computability Theory.* Elsevier; 1999.
22. Herken R, ed. *The Universal Turing Machine: A Half-Century Survey.* Oxford University Press; 1988.
23. Hodges A. *Alan Turing: The Enigma.* Burnett Books and Simon and Schuster; 1983.

24. Hopcroft JE, Ullman JD. *Introduction to Automata Theory, Languages, and Computation.* Addison-Wesley; 1979. Third edition (with Rajeev Motwani), 2006.

25. Immerman N. *Descriptive Complexity.* Springer-Verlag; 1999.

26. Kleene SC. *Introduction to Metamathematics.* D. van Nostrand Co.; 1952. Ishi Press; 2009.

27. Odifreddi P. *Classical Recursion Theory: The Theory of Functions and Sets of Natural Numbers.* North-Holland; 1989.

28. Post EL. Finite combinatory processes. Formulation I. *J Symbolic Logic.* 1936;1:103–105.

29. Sipser M. *Introduction to the Theory of Computation.* PWS; 1997. Second edition, 2005.

30. Soare RI. *Recursively Enumerable Sets and Degrees: A Study of Computable Functions and Computably Generated Sets.* Springer-Verlag; 1987.

31. Whitemore H. *Breaking the Code.* Amber Lane Press; 1987.

Index

Printed and bound by CPI Group (UK) Ltd, Croydon, CR0 4YY

03/10/2024

01040416-0016